# WHERE THE
# RIVER
# FLOWS

# WHERE THE
# RIVER
# FLOWS

SCIENTIFIC REFLECTIONS
ON EARTH'S WATERWAYS

*Sean W. Fleming*

PRINCETON UNIVERSITY PRESS

PRINCETON AND OXFORD

Copyright © 2017 by Princeton University Press
Published by Princeton University Press, 41 William Street, Princeton,
New Jersey 08540
In the United Kingdom: Princeton University Press, 6 Oxford Street, Woodstock,
Oxfordshire OX20 1TR

press.princeton.edu

Jacket art courtesy of Shutterstock

Library of Congress Cataloging-in-Publication Data

Names: Fleming, Sean W., author.
Title: Where the river flows : scientific reflections on Earth's waterways / Sean W.
Fleming.
Description: Princeton, New Jersey : Princeton University Press, [2017] | Includes
bibliographical references and index.
Identifiers: LCCN 2016028571 | ISBN 9780691168784 (hardcover : alk. paper)
Subjects: LCSH: Rivers. | Hydrology.
Classification: LCC GB1201.7 .F54 2017 | DDC 551.48/3--dc23 LC record available
at https://lccn.loc.gov/2016028571

British Library Cataloging-in-Publication Data is available

This book has been composed in Minion Pro

Printed on acid-free paper. ∞

Printed in the United States of America

10 9 8 7 6 5 4 3 2 1

The face of the water, in time, became a wonderful book . . .
which told its mind to me without reserve, delivering its
most cherished secrets as clearly as if it uttered them with
a voice. And it was not a book to be read once and thrown
aside, for it had a new story to tell every day.
—Mark Twain, *Old Times on the Mississippi*


Nature is the art of God.
—Sir Thomas Browne

# CONTENTS

# ACKNOWLEDGMENTS

This book would not have happened without the support, both direct and indirect, of many people. First and foremost, I must thank my editor at Princeton University Press, Eric Henney, for his guidance, insight, and faith in this project. Additionally, others at PUP have given invaluable support throughout the publication process, and several anonymous external reviewers provided helpful suggestions on earlier versions of the manuscript. Various mentors over the years have each taught me something important about how to be a scientist, including Garry Clarke, Don Russell, Anne Trehu, Roy Haggerty, Peter Clark, Marsh Lavenue, Paul Whitfield, Roland Stull, William Hsieh, Dan Moore, Julia Jones, and Chris Doyle. Many other colleagues—too many to list here—have also been invaluable in shaping my view of the practice of geophysics. I deeply appreciate the thoughts and tips on the process of publishing a general-audience book provided by Susan Danard, Bob Nixon, Robert Mackwood, and Shawn Marshall. Both a sympathetic ear and valuable feedback were provided by Sherry Mikkelson and Edgar Flores. And of course, special mention goes to my wife, Kristine, for her patience and proofreading, and to my entire family, including my mother Alice and sister Mieke, for their web of support. My early interest in science was spurred on by my father, Bill, who took me to the planetarium as a child for my first view of Saturn's rings through a telescope set up outside; and by my brother-in-law, Manjit, who as a young man brought electronics parts and projects home for me when he and my big sister looked after me. Thanks also to my father-in-law, Roger Penn, for use of the beach house as a writer's retreat to polish off the revisions. And of course, we can't neglect to mention our furry cheerleaders: Hank, Peanut, Mackenzie, Maggie May, and Mr. Harrison.

# WHERE THE
## RIVER
# FLOWS

# 1

## INTRODUCTION

Every culture recognizes that water is life, that rivers are the veins and arteries of the world. Rivers figure prominently in a Haida creation story, handed down by word of mouth among the First Nations of coastal British Columbia for innumerable generations, captured here by Martine Reid:

> As the Raven flew away, gasping to restore his breath and recover from his fright, drops of fresh water trickled down from his bill. These drops, falling on the mainland, became lakes and ponds of fresh water. From these, streams and rivers flowed in all directions.

Norman Maclean wrote semiautobiographically of his childhood home in Montana:

> Eventually, all things merge into one, and a river runs through it. The river was cut by the world's great flood and runs over rocks from the basement of time. On some of the rocks are timeless raindrops. Under the rocks are the words, and some of the words are theirs. I am haunted by waters.

These examples come from western North America. But one could point to any nation, any language, any creed, and find other stories similarly demonstrating a clear understanding of how powerfully important rivers are to us at every level.

Much of that significance is very practical. Rivers provide our drinking water. They give us food by irrigating our crops and watering our livestock and serving as the homes or birthing places for fish and wildlife. Rivers keep the lights on and commerce in motion by driving hydroelectric turbines. Their water is key to virtually any industrial process, from building cars to manufacturing computers. Rivers have served as transportation pathways for much of human history, and continue to do so to this day: the Danube still carries freight across central and eastern Europe, much as the Romans used it two millennia ago. It is no coincidence that a nation's GDP is linked to its water supply availability. Virtually every major city worldwide is built on a river, and rivers often delineate our national borders. Their valleys define where we are from, our homes, our nicknames, our industries, our animals, our culture, our booze, our sports teams, our place of mind. California alone provides a host of examples: the Valley Girl pop culture of the 1980s refers to the San Fernando Valley, which carries the Los Angeles River; Silicon Valley technology derives its name from the Santa Clara Valley, through which Coyote Creek flows; and the wines of Napa and Sonoma come from their respectively named river valleys. And what rivers bring, rivers can also take away; water is life, but it can also be death. We choose to build on floodplains, and then rivers' waters sweep away our homes and even our lives. Floods are the most expensive type of natural disaster in the United States. Drought is subtler but even more dangerous. It is reported to cause more human suffering globally than cyclones, earthquakes, and floods combined. And according to a United Nations report, deprivation in adequate clean water claims—through the associated disease alone—more lives than any war claims through guns. Developing a broad understanding of these issues has never been more important. Population growth and economic expansion are taking their toll on our watersheds, by increasing the demand we place

1.1. Aftermath of the 2015 Memorial Day flood in Texas. Photo: US National Weather Service.

on them for simultaneously providing both clean water and a dumping ground for waste, by likely affecting the climate that drives river flows, and by interrupting the natural landscape processes that regulate river levels.

How do we approach this? Where do we start? Stand on the bank of a river, and you'll notice one thing immediately: the water isn't at a standstill, but flowing by. It came from somewhere, and it's on its way somewhere else. At this moment, you are watching one part of the planetary-scale water cycle, the fundamental conceptual framework on which all of hydrology is based. Figure 1.2 illustrates some of these ideas. Ours is essentially an ocean planet: roughly three-quarters of Earth is covered by water, almost all of which is sea (salt) water. Freshwater constitutes the remaining 3.5% of the total planetary water store. And of that, more than half is locked up in glacial ice—mainly the gigantic Greenland and Antarctic ice sheets, but also alpine glaciers and ice fields atop various mountain ranges across the world. Groundwater is the second-largest component of the world's freshwater reserves. This is water stored underground in aquifers, which consist of the innumerable empty spaces between sand grains, for example. Rivers, on the other hand, hold only about 0.006% of Earth's freshwater resources—a negligible proportion of the total amount of water on the planet. How is it, then, that rivers are front and foremost in our minds? Is it just an error of perception? Not quite. The importance of rivers lies in the fact that everything is dynamic, linked together in a big loop, and rivers are the interstate highways in that global cycle. Rivers don't store much water at any given moment in time, but they move a great deal of it, making it available for drinking water and to drive turbines for hydroelectric power generation and the myriad other functions it serves.

Let's start our initial exploration of this cycle with the oceans. Evaporation under the sun's warmth moves water from the ocean surface upward into the sky, leaving behind the salt, a little like how a still concentrates alcohol to make whiskey. The atmosphere then moves that water around the world, through gigantic circulation cells and hemispheric-scale wind patterns— basically, the weather. Like rivers, the atmosphere acts mainly as a highway for water. For all the mighty thunderstorms and hurricanes we see on the nightly news, relative to other planetary stores the atmosphere actually contains very little water at any given time. Atmospheric water eventually falls to the land surface as rain and snow. It then undergoes all sorts of fascinating and important interactions, and we'll talk in detail about many of these. For now, just note that much of the rain and snowmelt eventually makes its way to rivers, which transport the water back to the oceans again—completing the cycle.

While this planetary water cycle forms the core concept behind all hydrologic studies, there are many aspects to hydrology and different takes on the subject, reflecting the diverse array

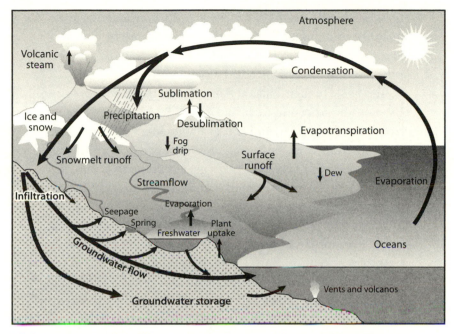

**1.2.** Many processes drive the movement and storage of freshwater. This intricately interlocked assemblage of processes is known as the hydrologic, or water, cycle. Image: U.S. Dept. of the Interior; U.S. Geological Survey; John Evans, Howard Perlman, USGS. http://ga.water.usgs.gov/edu/watercycle.html.

of implications that rivers have. For instance, civil engineers will often be concerned with understanding extreme river flow rates for designing bridges or delineating flood hazard zones. Forestry scientists are interested in the ways that forests, and different forestry practices, affect river flows and water quality. Agricultural and soil scientists study irrigation engineering and the impacts of agriculture on runoff. Atmospheric scientists often consider the potential flood impacts of their weather forecasts. Water supply managers are interested in overall river flow volumes and how these vary from year to year, especially periods of droughts. Oceanographers and life scientists occasionally make significant forays into the field due to its profound impacts on coastal oceans, and aquatic habitat quantity and quality. Geoscientists provide insight on topics like groundwater resources, water pollution, and the way that rivers change their paths, sometimes destructively.

This book, though, looks at rivers through the lens of physics. We won't be strict and exclusive in that emphasis—key topics from geology, biology, and chemistry, among others, will also be brought to bear. But we'll always return to our physics-inspired view of rivers. Our main goal in doing so is to shed a little light on the rivers that sustain our world. I aim to do that in a way that makes sense to you, regardless of how little, or for that matter how much, you might know right now about either rivers or physics. But I'm also aiming to communicate a view that's something more than just an assemblage of facts and explanations about the mathematical tools we use to forecast floods, or how greenhouse gases and climate change might affect water supplies, or how satellites monitor the glaciers and snowpack that power large rivers on every continent, or how urbanization impacts the ecological value of watersheds. We'll broach all those subjects, and many more. Yet across these individual topics, two larger threads will be interwoven: how looking at something from a fresh perspective can engender new types of understanding, and the interconnectedness of the environment and, indeed, of ideas.

We can consider this first notion—the value of looking at something familiar in a new way—at a couple of levels. One could be called, for lack of a better word, professional. The study of how water moves underneath, across, and above Earth is called hydrology, and the scientists who study these things are called hydrologists. Although basic physics has always had a strong bearing on hydrology, it probably wouldn't occur to most folks to regard hydrology as a branch of physics. Yet we take exactly that philosophy here. We can explore this idea by juxtaposing the geophysical study of rivers alongside the astrophysical study of the night sky. Of course, there are some obvious differences: while one looks up to the stars, the other looks to the creek in your backyard; one searches billions of light-years away, over expanses so vast that time and space merge, while the other offers a refreshing dip on a hot summer's day and then abruptly changes its mind and wipes out your house as its waters rise. But deep likenesses are also evident. Both apply similar methods to the analysis of uncontrolled natural experiments. Both can appeal convincingly to sophisticated quantitative tools like nonlinear differential equations and statistical mechanics and information theory, all of which we'll touch on later. Both seek to better understand the complex dynamics of systems we can control little if at all, not some carefully designed and tightly managed laboratory experiment but the very world around us, challenging us to tease the signal from the noise. Both peer into the heart of nature, in search of a flash of revelation. We posit here that reimagining hydrology as a sister science to, say, astrophysics, adds value by suggesting new ways of tackling some very difficult and very important questions about our natural environment, the ways we affect it, and the ways it affects us.

But there's also a much wider component to this idea of examining with a fresh eye something we think we know well. It's occasionally suggested that by attempting to reduce the universe to a rational, compact, fact-based explanation, the natural and social sciences can in a sense drain the joy out of our experience of the world, amounting to a sort of psychic vandalism. And at its most narrow-minded, science may do exactly that. Love, for instance, is a lot less intriguing if you choose to believe it's just a specific set of chemical reactions in your brain, and dismissing spiritual experience as a socially organized delusion is a condescending insult to cultures world-wide. At its best, though, science is just the opposite of this: an act of creativity and imagination, which adds new layers of meaning and significance to our familiar, everyday experience.

I find that the day-to-day change in the water level of a river I drive past on the way to work every morning, which might otherwise be so mundane I wouldn't even notice it, transforms into something magical when I think of the fractal dimensionality of this variability, how the tiniest hour-to-hour changes in water flow after a rain shower are linked across a vast continuum of time to the greatest variations in planetary climate and water cycles as Earth's orbit across the solar system slowly wobbles. A dramatic view of a river meandering across a desert landscape of red sand and sagebrush at twilight is made even richer by being able to look even deeper into this scene and recognize some of the levels of causality and complexity that contributed to it, from the rise of the mountains in its headwaters as a continental plate split apart over millions of years, to the way that the river shifts its channel when a thunderstorm descends from the skies to deliver a flash flood. And being able to contextualize the relationship among weather, rivers, and spawning salmon in terms of the same mathematical communications theory that drives the modern information age adds new depth to our intuitive understanding of an intricately intertwined natural world.

That brings us to the second thread that runs throughout this book: the notion of interconnectedness. Again, we can look at this at a couple of levels. The most basic and literal, though by no means simple, level is in terms of how rivers actually work. A river isn't just a single thing, but a complicated and interlinked collection of things, from the changing climate overhead, to the forests and snowfields upstream, to the thirsty farms and factories and cities downstream, not to mention the ecosystems of the watershed—that is, the area draining to any particular point on a river. And all of this can vary radically from one place to the next: in another part of the world, the aforementioned forests might be replaced by tropical jungles, or dry grasslands, or arctic tundra. So too for the hydrological sciences, which span all these topics and more, and need to accommodate tremendous regional differences between river basins. To understand rivers, then, we need to understand a lot of system components and how they all interact with one another.

At another level, though, we'll also spend a good deal of time in this book examining the broader interconnectedness of ideas. The tools of physics and mathematics that we can use to learn something about how El Niño affects river flows are also used to study deep-space phenomena like pulsars. The numerical methods that illuminate and quantify how wetlands impart a kind of inertia to stream water level variations are the same ideas and techniques used in stock market analysis. Looking at the mathematical and computational wherewithal that's brought to bear on flood forecasting also gives us a window into the world of artificial intelligence. And that mathematical theory of communication we mentioned earlier as a way of studying how weather, rivers, and ecosystems interact? Physicists are using these concepts to study black holes and, indeed, the information processing capacity of the universe itself. This universality of ideas, and especially mathematical ideas, is truly amazing, and we'll return to it again and again.

In summary, looking at hydrology in this new light, as not just a pragmatic exercise in applied science and engineering but instead a terrestrial equivalent to more ethereal disciplines like astrophysics, brings to mind exciting and even exotic ways of imagining questions about rivers that we'll explore here. Can rivers remember? Does a river choose its own course, or is it told where to flow? How do clouds talk to fish, and how much do they have to say? What does artificial life have to do with devastating floods and landslides? How do variations in Earth's orbit affect the alpine "water towers" that sustain much of the world's population? What is a digital rainbow, and how does it help us understand the way that rivers respond to climate variations? And what might be the greatest buried treasure of all? Drawing on examples from all over the world, and written to all audiences, this collection of short, light, illustrated chapters shows how the methods of physics, mathematics, and statistics can be used to answer these questions and more. Along the way, we'll explore some surprising connections: between a good-luck charm and El Niño, black holes and rain gauges, the history of life and the geology that helps determine where rivers flow, Wall Street derivative pricing models and river forecast systems—even between James Bond and groundwater wells. And in using this exploration of connections and explanation by analogy to develop a full-blooded, intuitive understanding of these topics, some of which (like feedback loops) are fundamental and pervasive throughout both the natural and artificial worlds, we'll see not only how the cosmopolitan intellectual realm of physics informs our view of rivers, but also how rivers can inform our view of physics and of the world as a whole.

# 2

## WHY RIVERS ARE WHERE THEY ARE

Perhaps the most basic question we could ask about a river regards its origin and placement: why is there a river here? This also happens to be one of the more challenging questions about a river to answer, because each watershed has its own unique history and personality. On the other hand, the laws of nature don't change from one watershed to the next, so it's a manageable task to trace out some broad generalities about why rivers wind up being where they are. In doing so, we'll have an opportunity to explore a wide variety of topics, ranging from feedback loops to climate changes to how the Cold War accidentally revolutionized Earth science. We'll conclude with a brief tour of a river in the Pacific Northwest and show how similar ideas work for other rivers worldwide, from the mighty Asian rivers flowing from the Tibetan Plateau, to the desert canyons of the American Southwest, to the Seine in the heart of Paris.

Let's broach the subject of how rivers set their courses by posing a narrower question. Standing at a lookout and peering down into a river canyon, you might ask: is the river there because of the canyon, or is the canyon there because of the river? In other words, did the river's turbulent flows cut down through the rocks over time to form the canyon, or was there a valley there to begin with, and the river simply followed it? Which came first, the river or the canyon?

This chicken-and-egg problem can be viewed as something called a positive feedback loop. The short and incomplete answer is that the canyon came first—sort of. Rivers often carve deep canyons that would not otherwise be there, but for there to be a river at that spot, there needs to be some kind of valley to start off with. Put another way, water flows downhill, so to get the focused, channelized flow required to form a stream, there first needs to be a depression in the landscape. This low point in the land might have originally been little more than a shallow topographical hollow. However, if the water flows with any significant speed, and the soil and rock are the least bit susceptible to erosion, then over time the channel will cut downward—and often across, as well as up-valley, capturing a bigger drainage area for itself. So the river carves out a larger and larger niche for itself in the landscape, creating or at least deepening the canyon. That is, a low point in the land gives a river, which gives a still-lower point of land, and so forth. Hiking or mountain biking trails give a close analogy. When people walk or ride on a path, they gradually erode a track. Then people go looking for tracks to walk or ride on, and doing so further erodes the track. This kind of positive reinforcement, whereby the effect of some cause becomes itself a cause of the same effect, is called a positive feedback loop.

Another geophysical example of a positive feedback loop is the melting of ice sheets. Albedo is the fraction of incident solar radiation (sunlight, more or less) that gets reflected by the surface of some landscape element, like a tree or a parking lot or a rock or a snowbank. The remainder gets absorbed, warming up the surface. Rock and soil and forests typically have pretty low albedos, meaning they absorb a lot of the sun's heat, and snow and ice have high albedos. That's why you risk snow blindness if you don't wear sunglasses while traversing a glacier or skiing on your local hill, even if it's not an especially sunny day: your eyes receive not only the sunlight from above, but also the reflected sunlight from below. And that reflected energy is energy that doesn't go toward warming up and melting the ice or snow. In a landscape widely covered by glaciers, the overall albedo is generally high. But say we have a long-term rise in temperature. It might be the result of natural climate changes like the variations in Earth's orbit that drove the ice ages, for instance, or a human-induced increase in greenhouse gas concentrations, caused by fossil fuel combustion and deforestation associated with population growth and economic expansion. We'll examine these important topics in more detail later. Over time, glaciers start melting away under higher temperatures, the proportion of the landscape covered by ice and snow decreases, and that covered by rock and soil and advancing vegetation increases. Consequently, the albedo of the landscape as a whole goes down. So, a greater proportion of the incident sunlight goes to warming the surface of the landscape, rather than just being reflected back up. This accelerates the melting and recession of ice sheets, and that in turn further accelerates the reduction in overall albedo, and so on.

The amazing ubiquity of feedbacks makes them a concept worth emphasizing. A perhaps more familiar example of a positive feedback loop is the screech of guitar-amp feedback: the electronics have some noise, which is broadcast by the amplifier as hiss or static, which is captured by guitar pickups (essentially microphones) that feed the noise back to the amp, which boost it and send it out on top of the original noise, and so forth into what can very quickly accelerate into a deafening screech. And when feedback loops are applied to engineering design, what you get is a big part of the field of cybernetics. This term may bring to mind the Terminator movies with their fictional Cyberdyne Systems and robot wars, but an example we're all familiar with is simply baking something in the oven. You set the oven to, say, 350°F and wait for the light to turn off so you can put in your cookies, and after you put them in, you can be confident that it'll stay around the right temperature. The oven has heating elements, a temperature sensor, and a control circuit. After you switch on the oven and set the dial, the oven turns on the heating elements and then monitors the temperature. When the sensed temperature gets to 350°F, the oven knows to turn off the elements. But it will also turn them back on again if it senses the temperature dropping too far below 350°F. This simple goal-seeking behavior (quite common in engineering applications, other examples being cruise control and antilock braking in your car) amounts to a negative feedback loop, in which the effect of some cause becomes in turn a cause of the opposite effect: excessive temperature causes the heating element to turn off and decrease the temperature, and insufficient temperature causes the heating element to turn on and increase the temperature.

So there we have the positive feedback loop that can exist between rivers and canyons. That's a start, but not every river flows through a canyon. And what about that initial topographic depression in the first place, big or small, the seed wherefrom the river grew, the reason why the river is where it is? Its nature and origin can vary greatly from one river to the next, and the specific answer can also vary depending on just how far back in time and how far up the causal chain you want to go. In virtually all cases, though, it's geological in nature.

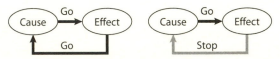

2.1. Left: In a positive feedback loop, the effect reinforces the initial cause, increasing the effect, increasing the cause, and so forth. It consists of self-reinforcing dynamics that tend to spin out of control. Right: In a negative feedback loop, the effect counteracts the initial cause, bringing the effect to an end. Negative feedbacks are characteristic of self-regulating systems that tend to stay on course.

A common example is the glacier-carved valley. These are ubiquitous in mountain regions that were extensively glaciated during the ice ages—a geologic time frame called the Pleistocene epoch, which ended roughly 12,000 years ago. The postglacial time frame, during which rivers started flowing again in these glacier-carved landscapes, is known as the Holocene. One may wonder why scientists invent geological time frames with such odd and obscure names and that don't start and end at nice, neat times—say, multiples of ten million years. The geological timescale was invented before physicists and chemists like Ernest Rutherford and Bertram Boltwood came up with radiometric techniques for precise dating. These are based on the decay rates of naturally occurring radioactive isotopes and the measured relative abundances of parent and daughter products in some particular rock sample; carbon-14 dating is a familiar example, although its short half-life is usually more appropriate to archaeological rather than geological dating. But it's not just a question of tradition. It's comparable to why people refer to the "Renaissance," as opposed to listing a specific set of dates: it clearly and concisely specifies a period of time having very particular characteristics. In the absence of precise numerical dating methods, early geologists classified Earth's history on the basis of the story told by the rocks. A key element of that storyline is the progressive evolution of life, which didn't care at all about adhering to a schedule laid out in multiples of ten million years. The age of the dinosaurs, for example, spanned about sixty-five million years ago to 248 million years ago, and it's convenient and meaningful to refer to this time frame as "the Mesozoic," as inferred from the geologically preserved fossil record. The history of rocks is in many ways the history of life, and indeed of everything associated with either, including global climate. So it makes sense to parse the geological timescale in this way. And some suggest we're now in a new geologic time frame called the Anthropocene, dominated by the effects of mankind. Later in this chapter, we'll see some examples of the imprint that the Anthropocene is leaving on the paths taken by today's rivers.

The landscape back at the start of the Holocene, though, was in many parts of the world largely defined by the preceding glaciation—actually, a whole series of glaciations during the foregoing Pleistocene, a topic we'll return to in a later chapter. This is true from much of North America to much of Europe, from parts of South America to parts of India, China, and Russia. In mountainous regions, the glaciers carved out a virtually infinite number of valleys. Some of these were buried under ice kilometers thick. As the glaciers began to recede, these valleys were the topographic low points in the landscape that in turn became home to newly born (or reborn) rivers. Things have generally remained that way to this day. Depending on local climate today, the head of the valley may or may not still house a glacier and its abundant summer melt water, but one way or another, the rivers are still there, fueled by some potential combination of glacial melt, snow melt, and rainfall. Going back further in time and up the causal chain, one might wonder why the glaciers carved a valley (ultimately birthing a river) in some particular spot—why is the valley here and the flanking mountains there, rather than vice versa? Reasons could be many, and may vary greatly from one valley to the next. In general, it has to do with the internal dynamics of the Pleistocene ice sheet, the comparative susceptibility of the rocks to glacial erosion in different areas, and the presence or absence of some degree of topographic low prior to the arrival of the glacier—potentially including a much older preglaciation river valley.

There are all sorts of other external forces that can act, either alternatively or additionally, to create the overall fabric and texture of the landscape, in turn forming something of a template for an emerging river drainage network. Many of these are related, directly or indirectly, to plate tectonics. Geology doesn't have an overarching Theory of Everything, nor does it really aim to in the way that physics does, but plate tectonics comes pretty close. The basic notion is that Earth's upper few kilometers consist of a number of relatively rigid plates collectively forming what's called the lithosphere, which move over a thick semimolten layer called the asthenosphere. The plates by and large each travel in a different direction, and the boundaries between them are where a lot of the action happens. Although it certainly doesn't explain everything, it does indeed explain a great deal, including global patterns of earthquakes, volcanoes, and mountain chains, not to mention the overall configuration of the continents. The shapes, locations, and numbers of plates and their boundaries all gradually change over many millions of years of geologic time, and the plate tectonic story is imprinted on the rocks and the landscapes we see today. Considering its central role in our collective planetary self-understanding, not to mention its associated array of spectacularly violent natural hazards, plate tectonic theory emerged only very recently. Although Alfred Wegener proposed continental drift in the late nineteenth century, definitive evidence and a clear mechanism for it, along with a wide-ranging theory of plate tectonics, had to wait until the 1960s. It's worth pointing out that, even then, it arose in large part as an indirect consequence of the technologies and needs generated by World War II and the ensuing Cold War, in particular comprehensive and detailed mapping of the world's ocean floors to facilitate such things as submarine warfare. This serves as a reminder that it's not uncommon for a fundamental discovery in some subject to be made as a by-product of other endeavors, and similarly, the potentially huge practical value of seemingly obscure theoretical work is often impossible to specify or predict beforehand.

And why is plate tectonic theory important to understanding why rivers are where they are? Because water flows downhill, and tectonic forces create hills—or mountains. In fact, any

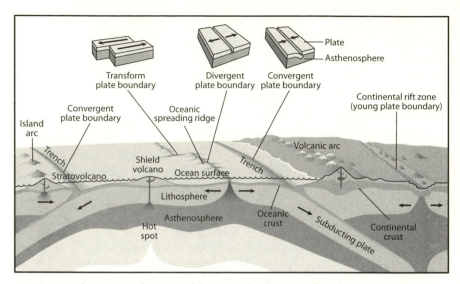

2.2. Tectonic theory. Two plates may slide past each other (transform boundary, like California's earthquake-prone San Andreas fault, separating the North American and Pacific plates), one may dive under the other (subduction zone, like where the Nazca and South American plates meet to form the Andes), they may collide (like the convergent boundary between the Indo-Australian and Eurasian plates, creating the Himalayas), or pull apart from each other (divergent boundary, like the Mid-Atlantic Ridge separating the North American and Eurasian plates in the middle of the Atlantic Ocean). Image: Cross section by José F. Vigil from *This Dynamic Planet*—a wall map produced jointly by the U.S. Geological Survey, the Smithsonian Institution, and the U.S. Naval Research Laboratory.

given landscape can be loosely viewed as the current local geological balance between tectonic mountain building and the forces of weathering and erosion. When two continental plates get slammed together, for instance, we see the same effect as two cars in a head-on collision: their hoods get crumpled downward and upward into jagged folds of sheet metal. This roughly describes how the Himalayas, Alps, Caucasus, and Rockies were formed. Or, when a thin but dense oceanic plate gets forced (subducted) beneath a thick but lighter continental plate, rock melts at depth and then rises, forming a volcanic mountain chain (a "volcanic arc") on the overlying plate.

Examples of this kind of mountain range include the Andes, which form the backbone of South America and the source of the Amazon; the snow-capped Cascades, spanning northern California and the Pacific Northwest, spawning a multitude of streams and rivers; and such well-known individual volcanoes as Mount Fuji in Japan, within sight of Tokyo, and Vesuvius in Italy, an eruption of which destroyed the Roman city of Pompeii. Weathering and erosion, in turn, wear those mountains down. Erosion by rivers, called fluvial erosion, may locally incise valleys, steepening the topography instead of flattening it—that is, forming the canyons we opened this chapter with. However, the net long-term effect of the many different types of weathering and erosion is to shave off mountains and redistribute their contents to the plains, perhaps not unlike the way the taxman shaves a bunch off the top of your income and then redistributes those dollars. That takes time, though, and geologically young landscapes have mountainous terrain as their dominant overall characteristic—along with the rivers that flow in the valleys between those mountains, running from the peaks above, down to the plains below. In fact, one might say that, especially in mountain regions, rivers are placed in dips in the landscape created by plate tectonics.

So the initial depression in the landscape, generally guiding where a river later appears, is often the result of much broader geophysical forces, like the waxing and waning of ice ages over tens of thousands of years, or the tectonic raising and then erosional leveling of mountain ranges over many millions of years. But what about the location of the river at a finer scale—why is it right here, and not 10 meters to this side or 200 meters the other way? Larger geological constraints can again sometimes play an important role. For instance, a narrow, long fracture zone containing weak, easily eroded rock lying next to strong, more competent rocks on either side will often pin down the precise location of the stream, and perhaps the canyon it might carve out for itself in the future. And the patterns of underground water seepage toward a river—we'll return to groundwater and aquifers in a later chapter—can, in some cases, be such that the total contributing area to a watershed doesn't quite correspond to what you'd expect based on the topography of the land surface, with groundwater flowing right across what would seem to be drainage divides between adjacent watersheds.

But the exact path a river takes will very often be a matter in which the river itself has a strong say. A dynamic interplay of local erosion and local deposition, guided in large part by the river's flow, can play a dominant role. And these processes happen much more quickly than glaciation or tectonics. Let's begin by considering how rivers decide what their own beds and banks should look like. To see this how this works, we need to appreciate that the maximum size of materials transported by a river depends on the channel steepness and water depth. These two factors determine the shear stress exerted by the moving water on sediments lying on the streambed. Think of shear stress as the force exerted on a surface, in a direction parallel (or tangential) to that surface. As an approximate analogy, consider a car at a standstill on a gravel road. If you punch the gas, the spinning tires will exert immense shear stresses on the road surface, tearing it up and spraying gravel behind your car, *Dukes of Hazzard* style. If you instead ease on the gas, the car will begin inching forward without drama, as the tires are exerting little backward-directed force on the road surface, and the gravel stays where it is. In a river, the shear stress is provided by the force of moving water. The archetypal river—and to be sure, not all rivers are exactly like this—consists of steep, shallow tributaries in the upstream headwaters, and flatter, deeper sections in the downstream floodplains. As we move downstream, though, steepness generally decreases a lot more than depth increases. So the

2.3. Satellite image of the mouth of the Mississippi River and its sediment plume in the Gulf of Mexico. The blues musician, Muddy Waters, grew up here and apparently was nicknamed after the waters of Deer Creek, which flows through the delta. Due to the area's flatness, the river loses its ability to transport sediment carried from far upstream, and fine materials drop out of the water to create the muddy Mississippi Delta. Photo: Provided by the SeaWiFS Project, NASA/Goddard Space Flight Center, and ORBIMAGE.

shear stress, that is, the net ability of the river to move large objects, usually decreases downstream. This general tendency helps explain some common patterns you may have noticed. A stream will often be full of boulders in its steep mountain headwaters, where much of the available sediment is flushed downstream. Lower sections have more modest gradients and will often tend to have beds and banks made of sand and gravel, as these larger bits drop out of the stream due to lower shear stresses between the flowing water and the streambed. And far downstream, near the mouth, where typically the land is flat, the streambed and delta sediments will often be silty or muddy as these finest sediments finally drop out of the river.

Similar general principles also help determine the local-scale "morphology" of the drainage network—in other words, why the river is where it is and why it looks the way it does. In the headwaters of glacial rivers, for instance, rivers may cut downward to form canyons as we discussed above, but this is not the most usual case. Rather, the huge availability of glacial sediment, the high gradient and massive volumes of melt water during the summer freshet, and the steep-sided but relatively flat-floored cross-section typical of glacier-carved valleys (a characteristic U-shape as one looks up the valley) are such that coarse sediment is continually picked up, moved a bit, dropped, and moved again, leading to ever-shifting local topography. The net result is not a single well-defined stream channel, but rather a whole series of multiple channels that intertwine, separate, and recombine in what is called (due to the overall visual effect) a braided stream. Channel locations can vary quickly not only in space but additionally in time, with the channel pattern shifting from one melt season to the next. These braided channels tend to be very common and easy to spot in the sparsely vegetated headwaters of glacial rivers, but you can also see them in lots of other rivers, glacial or not.

Another very common channel morphology is the meandering river. This likely is the most widespread type globally. The meandering river is exactly what it sounds like: it meanders or wanders from side to side across the valley as you go downstream, giving a snake-like appearance in map view. It's not uncommon for these sideways meanders in the river to be bordered and limited by some broader geological feature, such as a bedrock ridge, but this is often not the case. By and large, the meanders—and thus the precise location of the river—are generated by the river itself, again as a result of local-scale patterns of erosion and deposition. And again, the meandering river is a dynamic environment. At outside bends in the river, currents are strong and directed somewhat toward the bank, eroding it. At inside bends, the bank is more sheltered, and the material that had just been eroded from outside bends is deposited, forming sand and gravel bars for instance. Consequently, material is systematically shifted about, and the meanders move over time, both down-valley and side-to-side across the valley. In effect, the river changes its own location. Sometimes a meander even gets cut off from the river in this way, forming what is called an oxbow lake.

As they tend to occur on flatter terrain, meandering rivers typically spill their banks on a fairly regular basis and the inundation may extend some distance away from the stream channel. This leads to the deposition and ongoing renewal of abundant and fertile soils on the floodplain. Aside from flattening the terrain further by smoothing over the bumps, this sedimentation also makes the floodplain a great place for agriculture. And that in turn makes it a great place for human settlement, which then leads people to build various structures for flood control, like dams and dikes. But as the overbank flooding ceases with the construction of such devices, so too does the geomorphological and biogeochemical refreshing activity associated with floods. Over time, the rich agricultural soils may deteriorate as a consequence. Furthermore, streamside (riparian) or wetland vegetation typically dwindles in such cases or is deliberately removed. This compromises rich streamside ecosystems. Additionally, the biogeochemical processes that occur in riparian areas are highly effective water purification systems, so that contaminated agricultural and urban runoff may often run more directly into the river when riparian vegetation and wetlands are lost. There are more direct economic and safety issues, as well. For instance, levee or dike systems are intended to contain high flows within the river channel itself. However, this results in greater water depths within the channel than would occur under natural conditions. If the flood flows are extreme or there is a flaw in the dike system, then the dike may be breached. But rather than the relatively gradual overtopping of the banks that would be encountered under many natural circumstances, the greater power of the channelized river may blast violently over and through the high levee, causing acute local damage to farms, homes, and whatever else may have been constructed there. In addition, as a general rule such control structures (apart from being pricey to build, inspect, and maintain) may lead directly or indirectly to losses of recreational and commercial fishing industry dollars, clean drinking water supplies, and quality of life.

But not all is dreary and depressing. Good planning, clever engineering, a bit of investment, and geophysical and ecological knowledge and the resulting ability to understand consequences and trace out options, can go a long way toward effectively preserving and rehabilitating riparian areas and floodplains, circumventing a lot of these problems. That is, one way or the other, ye shall reap what ye sow, as the old biblical saying goes. To this end, watershed rehabilitation efforts are ongoing worldwide. Perhaps a particularly good example is the "room for rivers" program in the Netherlands, intended for the dual purposes of restoring habitat and better handling flood flows. In some parts of the world, largely pristine watersheds still exist, so there we have the luxury of being able to preserve natural rivers rather than having to try to reconstruct them.

To see how some of the ideas we've talked about actually come together on the ground, let's take a brief headwaters-to-mouth tour of a river. We've noted that each river has a unique story, but that the underlying principles are universal. We can't consider every river in the world here, so we'll just pick a manageable and instructive one and then give examples of how the features we see in it appear in other rivers globally. While the details are specific to this watershed, we'll show that a similar story can be constructed for any river, including the river nearest you—wherever that may be.

First, a few words of context about our pet watershed. The headwaters of the Capilano River lie in the steep rain forest wilderness of the Coast Mountains on the southwest coast of British Columbia. Downstream, it forms the municipal boundary between two suburbs, and ultimately flows into the sea within view of downtown Vancouver's skyscrapers—specifically, into Burrard Inlet, an arm of the transboundary Salish Sea, which reaches south to Seattle and connects to the Pacific Ocean. Damming first occurred in the nineteenth century. Today, reservoirs on the Capilano and nearby Seymour Rivers together provide about 70% of the drinking water for the Vancouver area. Human interventions have deeply, and perhaps irreparably, compromised the river's ecological integrity. However, a strong hatchery program has maintained a reasonable salmon and steelhead presence, maintaining both traditional First Nations and recreational fisheries. Given the river's proximity to a potential urban angling population of perhaps three million people, it's unsurprising that the Capilano's fish have grown streetwise. In spawning season, one can watch gorgeous salmon leap right past the offerings of dozens of skilled (and of course some not-so-skilled) anglers. Nevertheless, seals, otters, and eagles line up alongside the anglers near the mouth in early autumn to sample a measure of the river's bounty. An analogous tension between nature and urbanization, reflected in the health or lack thereof of an area's rivers, can be found in almost any city worldwide.

We'll start our tour of this watershed a little way upriver, at Cleveland Dam, looking upstream across the reservoir of Capilano Lake and into the heart of the Coast Mountains (see fig. 2.4a). Rising from sea level to a couple thousand meters' elevation in just a few kilometers, these mountains force moist incoming air from the Pacific sharply upward. A similar effect occurs along the entire western margin of North America, from the Sierra Nevada in California to the Wrangells in Alaska, and indeed at mountain ranges worldwide. The air consequently cools, and water condenses out as snow and rain—and lots of

2.4. Mountain headwaters. (a) Capilano River, looking north from Cleveland Dam. Photo: S. W. Fleming. (b) Plate tectonics and glaciers similarly helped fix the locations of the upstream parts of many river channels worldwide. Collision of the Indo-Australian and Eurasian plates caused Earth's crust to crumple up, raise the Tibetan Plateau, and form the Himalayas and other great mountain ranges. The Yangtze Basin is the largest river by flow volume in Asia and the third-longest in the world, housing one-third of China's population. Shanghai was built at its mouth, and its headwaters lie in the glaciers of the Tanggula Mountains, on the edge of the Tibetan Plateau, shown here. Photo: Charlie Fong / Wikimedia Commons.

it, plenty to support a host of rivers here, just as the snows of the Rockies, Alps, and Himalayas feed many of the planet's greatest rivers and provide water supplies to billions of people. This process of adiabatic cooling and orographic precipitation is so important that we'll come back to it in considerable detail in a later chapter.

Why are these magnificent mountains here, both controlling the local climate and placing the headwaters of the Capilano River where they are? As for virtually all mountains, plate tectonics play the starring role. The Pacific Northwest is the site of a subduction zone, and has been for roughly 200 million years or so at this particular location. As we briefly discussed earlier in this chapter, a subduction zone is where an oceanic plate slides—in fits and starts, producing massive earthquakes along the way—beneath a continental plate. But sometimes the descending oceanic plate has larger bits, such as volcanic island chains, on top of it. A particularly large set of such bits collectively forms a narrow but long strip, spanning what's now southern British Columbia all the way to the Alaskan Panhandle. Carried atop a subducting oceanic plate, it rammed into the previous shoreline over an extended period about 100 to 150 million years ago. The heat and pressure associated with this subduction and collision produced a new volcanic mountain chain, analogous to the Cascades or Andes for instance, more or less where the Coast Mountains are now. The roots of these volcanoes consisted of granites, formed about 100 million years ago from magma that slowly cooled and solidified deep under-ground. The resulting giant, buried rock masses are collectively about 2,000 kilometers long and are known by geologists as the Coast Plutonic Complex. Ongoing compression between the tectonic plates lifted the land skyward, eroding away the overlying volcanoes, and exposing the underlying granites to form the ancestral Coast Mountains. Worldwide, granites formed underground and were then uplifted and exposed at the surface in a roughly similar way, from Mount Rushmore and its massive memorial sculptures of US presidents in South Dakota, to the Peninsula Granites of South Africa, acting as a base atop which the sandstones of Cape Town's distinctive Table Mountain stand like a pedestal. Roughly forty-five million years ago, those tectonic processes eased off, allowing the ancestral Coast Mountains to erode down to a series of hills. In fact, by about ten million years ago, the coastal range was sufficiently low that it no longer cast a rain shadow over the interior— which had earlier been (and is now again) a high desert. But during the late Miocene to early Pliocene, a geologically very recent five million years ago, the configuration and motions of the plates changed yet again. This caused renewed uplift of the Coast Plutonic Complex, again pushing skyward the Mesozoic granites. That uplift created the modern Coast Mountains, and continues today across much of the Pacific margin of North America, with upward movement in coastal British Columbia of as much as 4 millimeters per year in places.

The next major set of events shaping the above view into the Capilano's headwaters was the Pleistocene glaciations. Reflecting variations in Earth's orbit around the sun that we'll also discuss in more detail in a later chapter, the Cordilleran Ice Sheet coalesced from many individual glaciers in the mountains of western North America. Along the Pacific Coast, the most recent of these glaciations extended southward to roughly the present-day location of Seattle. It ended about 12,000 years ago, leaving the U-shaped glacial valley home to the upper Capilano River today, just as other ice sheets and glaciers carved out the paths followed by the headwaters of major rivers globally, ranging from the Danube of central and eastern Europe, to the Mekong of southeast Asia, to the Indus of Pakistan, India, and China.

Our next stop on the Capilano River tour is a fishing hole called Doctor's Pool, which marks the downstream end of a canyon (see fig. 2.5a). There's a cliff here that offers a concise visual story about the geologic history of the area. As mentioned above, long-ago tectonic uplift created the ancestral Coast Mountains, which in turn began to erode away. The materials that were eroded from these mountains were deposited by long-gone ancient rivers in a low-lying area called the Georgia Basin. This deposition began in the Cretaceous, about eighty million years ago, and continued for perhaps forty million years, into the Tertiary. Georgia Basin sedimentary rocks consist of conglomerates (gravel held together by finer sediments to form something that looks like concrete), sandstones, and still finer-grained rocks like mudstones. To get some feel for the corresponding time spans, dinosaurs went extinct at the end of the Cretaceous, and mammals began their evolutionary rise in earnest at the start of the Tertiary. So it took a while to put together the rocks of the Georgia Basin. The formation of sedimentary basins like this one drives much of the global economy, as these are where our fossil fuels were formed. Coal and oil are the remnants of plants buried by sediments and then, very slowly, chemically transformed by the weight of overlying rock and Earth's internal heat. After extraction at coal mines or oil wells, these geologically "cooked" plant remains provide energy when burned and are also used to produce a variety of materials, from nylon to body lotion. Like rivers, each sedimentary basin has its own history, but some common principles underlie all of them, and the oil wells of Texas, the North Sea, and Saudi Arabia all tap into sedimentary basins that are something like the Georgia Basin—itself a significant coal producer in the times of the British Empire.

Using its own language—in particular, by showing us the approximately eighty-million-year-old conglomerates of the Georgia Basin, lying immediately upon granites similar to those from which they were originally derived—this cliff on the

(a)

**(b)**

2.5. River-cut canyons. (a) The Capilano River at Doctor's Pool. (b) Such canyons are found worldwide wherever the conditions are right. As larger plate tectonic forces pushed the Colorado Plateau upward over about five million years, the Colorado River eroded downward through the rock to create the Grand Canyon. Photos: S. W. Fleming.

Capilano River tells us about the many other ancestral rivers that were here before it. But what does this tell us about the reasons for the river's location? The answer comes from asking another question: why can we see all this formerly buried geology in the first place? Because it was uplifted by plate tectonic forces (you can see the resultant tilting of the rock layers in fig. 2.5a) and then, through all of it, ran a vertical fault along which large adjacent blocks of rock slid in different directions. This, in turn, is why the Capilano River runs right here, and not

100 meters to the west of here or 300 meters to the east. The canyon is the result of a fault cutting through all these rocks. The fault forms a narrow zone of geologically weak material that is in turn highly susceptible to erosion by the river—more or less the type of positive feedback loop discussed at the start of this chapter.

For the most downstream stop on our brief tour of this river, let's consider the view looking upstream from a bridge (see fig. 2.6a) located roughly a kilometer from the mouth in a thoroughly urbanized area with single-family housing to the east and condominiums to the west. The evolution and present state of the lower Capilano has been strongly shaped by processes during and after the last ice age. A case in point is the postglacial development of the Capilano River's alluvial fan and marine delta. When the glaciers first retreated from here, sea level was a couple hundred meters above what it is presently, and marine waters flooded in. The land had been depressed by the weight of the ice sheet—Earth's crust actually sank farther down into the semimolten asthenosphere far below, not unlike the way a loaded freighter rides lower in the water. It subsequently took some time to rise back up when the ice receded, so that for a while it was flooded by the sea. But rise back up it did, and in fact, relative sea level dropped well below its current position for a spell, because while the crust had almost fully rebounded in this region, much of the world's water still remained locked up in massive ice caps elsewhere. Eventually, that ice melted off too, and sea level settled at roughly its present elevation. And through this whole time, the Capilano River carried its heavy load of rock and gravel and sand southward to the north shore of Burrard Inlet where it was deposited, forming about a six-kilometer-long stretch of the present-day shoreline of the inlet. Rivers and shorelines worldwide similarly bear the marks of varying sea levels.

Some of the most profound changes in the lower Capilano River, though—and a big part of why it flows exactly where it does today—occurred within the last century, as a direct and deliberate result of human intervention. For instance, until the 1930s or so, the Capilano had a substantial delta at its mouth, with multiple distributary channels (kind of like tributaries, but at the downstream end of a river instead—you can see examples of these in the satellite image of the Mississippi Delta earlier in this chapter). Indeed, looking at old maps and photos, the river appears to have had two main mouths, emptying into Burrard Inlet on either side of a low-lying island. The complex delta environment at the mouth of the Capilano was likely both ecologically rich and difficult to build things on, and it evidently was done away with by engineers around the time the iconic Lions Gate Bridge was constructed at this location.

Another powerful and deliberate change to the lower Capilano was the addition of riprap, shown in figure 2.6a. Riprap simply consists of large blocks of rock placed along a riverbank. The idea is to armor the banks against erosion and keep the river in its present path. The lower reach of the Capilano River is not constrained by bedrock geology like it is in the upstream canyon, instead meandering across its own floodplain. Recall from earlier in this chapter that such meanders tend to migrate over time, through a process of erosion in one spot and deposition in another, so that the river picks its own course and continually changes its mind about where exactly it wants to flow. Needless to say, that can be problematic to any built structures in the neighborhood, like homes and bridge foundations, hence the prevalence of riprap or similar "channelization" efforts along so many riverbanks worldwide. The downside is environmental loss; for example, replacement of streamside vegetation and side channels—important freshwater habitat—with a largely featureless and unsheltered

(a)

**2.6.** Channelization. (a) Lower section of the Capilano River, with riprap at the right-hand side. Photo: S. W. Fleming. (b) Many European rivers were straightened and walled in over the centuries, to the point that they're little more than canals, yet they offer a magic all their own. The Seine, shown here, remains the focal point of Paris. Indeed,

**(b)**

Notre Dame cathedral lies on an island in the river, and the arts quarter is known as the Left Bank, referring to the fact that it's on your left as you travel downstream on the Seine. Photo: Toni Frissell / Library of Congress / Wikimedia Commons / Public Domain.

embankment of angular boulders or simply a concrete wall. Still, along this and adjacent reaches of the Capilano, there's plenty of chances to see bald eagles, kingfishers, American dippers, and blue herons, not to mention salmon and steelhead supported by the upstream hatchery. In many ways, and like so many other rivers, especially in urban areas, the Capilano is no longer what it once was—but it's not beyond hope, and it has its own unique appeal, blemishes and all.

# 3

## HOW DO RIVERS REMEMBER?

Watersheds remember—culturally, geologically, ecologically—insofar as what happened before affects what happens now, and how things are now affects what is to come. A past culture of reverence or neglect will tend to bode well or ill for the future. Where the glaciers lay millennia ago helps determine where the river flows today. A strong or weak pink salmon run this year may promise strong or weak returns two years down the road, when the adults who were born this year come back to spawn in turn. Of course, such predictions are only probabilistic: unforeseen events happen, people's environmental attitudes change for the better or for the worse, rivers spill their banks and choose new routes to the ocean, open-sea salmon habitat conditions may be favorable or adverse. So there's also a more or less random element to it. The past influences but doesn't predetermine the present.

This is true of a river's flow as well. The single most fundamental and overarching aspect or measure of a river is its flow—how much water is coming down the channel—and how that varies over time and space. Streamflows influence just about everything else about a river, from habitat availability to bridge design, from water quality to water supplies, from flood hazards to white-water rafting conditions. And it requires only a moment's reflection on the behavior of the creeks and rivers you've encountered in everyday life to realize that their flows show a kind of memory. With only a few exceptions, flows generally don't change from a trickle to a flood from one hour to the next. Rather, if water levels were high yesterday, they're likely to again be high today; to put it another way, today the river remembers what the flows were like yesterday. In this chapter, we'll take a closer look at how this works. Doing so will not only give us a deeper appreciation of how rivers work but also an opportunity to explore a lot of other concepts—ranging from chaos theory to stock market volatility to artificial intelligence—concluding with some thoughts about objectivity, subjectivity, and progress in science.

The first thing to do is define what we mean by a time series. As the name implies, a time series is a series of measurements of something over time. For us, that "something" is river discharge or streamflow, usually measured in cubic meters of water per second, abbreviated $m^3/s$, $m^3s^{-1}$, or cms. What a flow of 1 $m^3/s$ means is that, if you're standing on the riverbank watching the river flow past, then a volume of water, equivalent to a cube 1 m wide by 1 m long by 1 m tall, will flow by you every second. Other units are possible too, ranging from the cubic feet per second used by the US Geological Survey and others, to the acre-feet per year still often used in reservoir studies, through to some really obscure or archaic ones like miner's inches. Let's get a feeling for what these river flow numbers mean and how much they can vary between watersheds. The average flow of the world's largest river—the Amazon—is roughly 200,000 $m^3/s$, equivalent to eighty Olympic-sized swimming pools full of water passing by each and every second. The numbers drop off pretty quickly from there. The Yangtze River, the largest in Asia and in the top ten globally, has an average streamflow of around 30,000 $m^3/s$. For the Willamette, a river in the rain forests of western Oregon that flows into the Columbia at Portland, the Hudson, which defines both the western bank of the island of

Manhattan and the New York–New Jersey state line, and the Murray-Darling, Australia's breadbasket and longest river, the numbers fall in the 600 to 900 $m^3/s$ range. And the average flow of Rio Chiquito, a small creek in the high desert of northern New Mexico, is only about 0.24 $m^3/s$. The actual values can be much higher or lower than that at any given point in time, however. And this, of course, brings us back to time series.

An example of a time series is a decade's worth of monthly average streamflow values for a particular basin, as shown in figure 3.1b for the Tigris River near its mouth in Iraq. The Tigris and its neighbor, the Euphrates, are perhaps the most significant rivers in all of human history, because in tandem they defined Mesopotamia. This region formed the eastern side of the Fertile Crescent, an arc-shaped area curving from the Nile River valley and delta in northeasternmost Africa across the present-day Middle East to the Persian Gulf. The Fertile Crescent was the first site of what we consider to be the most fundamental, defining inventions of human civilization: writing and agriculture. The Sumerian and Assyrian empires were based in Mesopotamia, and tradition holds that Abraham, the father of the Jewish, Christian, and Muslim faiths alike, hailed from here originally. The ancient city of Babylon was located on the banks of the Euphrates, not far from the modern city of Baghdad, which lies on the banks of the Tigris. The life-sustaining waters of these two desert rivers were key to Mesopotamia's central place in world history, and indeed, the word Mesopotamia derives from the Greek for "between rivers." The headwaters of the Tigris and Euphrates are located a short distance apart in the mountains of Turkey; the Tigris then flows southeast through Syria and Iraq, eventually joining the Euphrates again in a shared delta, where they flow together into the Persian Gulf, not far below where our flow measurements were taken. A sequence of monthly flow values does a great job of revealing the seasonal cycle, which is the largest

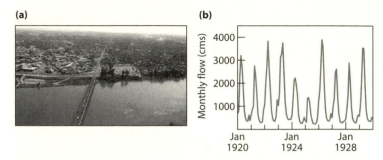

**3.1.** Monthly flow of the Tigris River using data from US National Center for Atmospheric Research. Photo: view from the air of the modern Iraqi city of Mosul, across the Tigris from the ancient city of Nineveh, from Sgt. Michael Bracken / PD-USGOV-MILITARY-ARMY / Wikimedia Commons / Public Domain.

type of variation that most rivers experience, and we can clearly see this for the Tigris time series shown here. The flows rise and fall like a heartbeat with a once-a-year pulse, peaking in spring from rainfall and snowmelt in the distant mountain headwaters, and dropping to a minimum around September.

But that seasonal heartbeat can be a little irregular for some rivers. Consider the Congo River (figure 3.2). In terms of flow volume, the Congo is the second-largest river in the world after the Amazon, and it drains over a tenth of the entire land mass of the African continent, with a watershed spanning no fewer than ten countries. Its headwaters lie in the East African Rift and adjacent areas, where the human species traces back its evolutionary history. An interesting feature of the Congo is that the equator passes through the basin, such that at any given time, it's winter in one part of this huge watershed and summer in the other. Complexities like these result in an average of two seasonal peaks per year at downstream locations like Kinshasa: a major peak around December, and a secondary one around May.

We're not restricted to monthly measurements, of course. Another example of a time series is 365 daily measurements

**(a)**  **(b)**

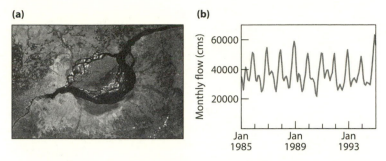

3.2. Monthly flow of the Congo River using data from US National Center for Atmospheric Research. Photo: satellite image of the capital cities of Brazzaville and Kinshasa, courtesy of the Earth Science and Remote Sensing Unit, NASA Johnson Space Center.

of streamflow for some particular river during some particular year, as shown in figure 3.3 for Boulder Creek, which flows from the Front Range of the Colorado Rocky Mountains down to the plains not far from Denver. Almost every creek is a tributary to another: runoff cascades from one creek to an incrementally larger one, and then to the next larger river, and the next after that, forming a great system of watersheds, one nested within the other. In Boulder Creek's case, its waters drain into the St. Vrain River, which then empties into the South Platte River, which subsequently joins with the North Platte; the Platte then empties into the Missouri, which flows into the great Mississippi and finally the Gulf of Mexico. That is, every great basin is a collection of many smaller basins, and the city of Boulder happens to derive much of its water supply from this one, running right through the middle of town. The city's forefathers, recognizing their position in the dry American West and the likely water resource implications of further settlement and population growth, showed great vision in purchasing the headwaters of the Boulder Creek watershed and ensuring a relatively stable municipal water source far into the future. One consequence of this choice is that Boulder is apparently the only city in the

**(a)**

**(b)**

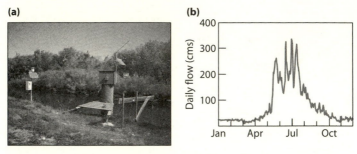

3.3. Daily flow measurements for Boulder Creek during 1993. Photo: equipment for flow measurement and real-time data telemetry at one of the stream gauges on Boulder Creek, from USGS.

United States that can claim to own a glacier—specifically, the small and unfortunately quickly receding Arapaho Glacier. By looking at a smaller "sampling interval"—the time between measurements, which for our Boulder Creek example is a day rather than a month—we can resolve finer-scale hydrological processes. In the time series graph of figure 3.3b, we can still clearly see the seasonal pattern, which is in this case a general yearly streamflow maximum around July or so, corresponding to snowmelt in the mountains. But now we can also see individual flow events corresponding to day-to-day weather variations and the watershed's response to them. Examples might include particularly warm spring days powering snowmelt, or perhaps summertime thunderstorms dropping buckets of rain. We'll return to some daily streamflow data very shortly, equipped with the basic physics of time series memory and an artificial intelligence technology.

Not only can we change the sampling interval, but we can also consider things other than flow rates. One of the most famous data sets in hydrology—in fact, perhaps the only famous dataset in hydrology—is the record of annual minimum water levels of the Nile River near Cairo. Spanning several centuries during medieval times (about AD 600–1400), these measurements were

**(a)**    **(b)**

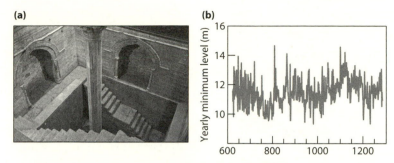

**3.4.** Annual minimum water level of the Nile River over about AD 600–1400. Photo: Nilometer at Rhoda, from Baldiri / Wikimedia Commons.

obtained by manually observing the yearly minimum elevation of the Nile waters in a large stone construction, illustrated in figure 3.4, called a Nilometer. The Nile is the world's longest, and perhaps longest-studied, river. Starting from headwaters many thousands of kilometers to the south, not far from the source of the Congo, the river flows north to the Mediterranean Sea, and its annual floods deposited the rich agricultural soils that supported the pharaohs of ancient Egypt—the western anchor of the Fertile Crescent mentioned above. As such, the river's annual behavior was observed with great interest. And it still is: studies of this Nilometer data set by Harold Edwin Hurst, a mid-twentieth-century physicist-turned-hydrologist, resulted in the discovery of what subsequently came to be known as the Hurst effect. This is a complicated sort of pattern in time, related to fractals, that has now been observed across virtually all of nature, ranging from ecology to electronics to medicine and even some social processes. We'll talk in more detail about that later in this book, but we'll make a note of it now because the process is an infinite version of the type of streamflow memory we'll be introducing shortly in this chapter.

It's easier to explain, and to understand, questions around memory in river flows if we can use a few symbols. Streamflow is usually represented using the letter $Q$, though what symbol is chosen is arbitrary so long as you clearly define what you mean by it. And because our time series consists of stream-flows measured at consecutive times, we can be a little more precise and denote it $Q_t$, where $t$ stands for time. So if we're considering the Boulder Creek daily time series shown above, for instance, then our first daily streamflow value is $Q_{t = 1}$, on the second day of our time series it's $Q_{t = 2}$, and so forth until we get to our last flow measurement, $Q_{t = N}$, where $N$ is just the number of data points we're working with (say, $N = 365$ for a year-long daily record).

To understand streamflow memory, we need to consider a type of equation called an iterated map, also known as a recursion or recurrence relationship. Iterated maps are a kind of mathematical model that describes the current value of whatever we're concerned about in terms of its previous values. Hence the "map" moniker: the equation gives us a map for how to get from the value of something at one time to its value at a later time. To reproduce a time series, the same map is reapplied over and over again, time after time. The notable trait of this kind of equation is that it can be really simple, yet produce an extraordinary diversity of (often very complex) behaviors that do a decent job of explaining the dynamics of a lot of complicated real-world systems.

An interesting example is the logistic map, $X_t = r\, X_{t - 1}$ $(1 - X_{t - 1})$. It's mainly used for describing biological population dynamics in simple, isolated systems governed by the balance between resource availability (such as food) and resource demand (as reflected in the size of the population, and therefore its need for things like food). In this case, $X_t$ is just a symbol representing the population size over time. $X = 0$ means a population of zero, and $X = 1$ means the maximum possible population—that is, the carrying capacity of the ecosystem, with any more than that resulting in a population collapse because resource demand exceeds resource supply. The symbol $r$ is what's called a model parameter, and it's simply a constant number. Note that by "constant," we mean only that $r$ is fixed for a given ecosystem (or petri dish, as the case might be). If we look at a different species of algae, say, in a different petri dish, $r$ might also be different, though it would in general be constant over time for that second petri dish. We can change the number we give to a parameter without changing the equation, but the particular number we give to this parameter can change the way that the equation as a whole behaves. The logistic map or closely related equations are also used to model other systems, such as long-term global oil supplies, and as a "toy" model to mathematically explore chaos theory. This brings us back to that interesting thing about iterated maps: although strictly speaking there's little or nothing in the above equation that you wouldn't see in junior high school algebra, it turns out that when we set the value of $r$ to 4, this equation produces deterministic chaos.

As it turns out, chaos theory has some fairly profound implications. Formally, deterministic chaos refers to an exponential sensitivity to initial conditions. Informally, it speaks to how a very small change in how things start leads to a huge difference in where they end up, and it explains a lot of things that look random but aren't. Such deterministic chaos is also referred to as the "butterfly effect," a phrase ascribed to the atmospheric scientist Edward Lorenz. Using a different set of equations than the logistic map described above, he invented modern chaos theory as a directly applicable physical science, although it was predated by parallel concepts in abstract mathematics and some other geophysical developments like Tsuneji Rikitake's dynamo model for reversals of Earth's magnetic field. Lorenz's interest was in numerical weather prediction, and the aforementioned butterfly effect refers to a semiapocryphal story about how a butterfly flapping its wings in Brazil might make the difference as to whether a tornado does or doesn't appear over Texas (though some attribute the term to the fact that a particular way of graphing the solutions to the Lorenz equations traces out a pattern that looks like a butterfly flapping its wings; others substitute China for Brazil; and so forth). A practical punch line is that deterministic chaos goes a long way toward explaining why accurate weather forecasts are so tough to make. But these mathematical ideas extend far beyond fluid mechanics and other applied physics problems like weather prediction, and have been invoked in the life sciences (including but not limited to the aforementioned logistic map) and even in the social sciences. One of the broader and more philosophically interesting points about chaos theory is that it lies in some sense between order and randomness, between determinism and free will. It implies that while strictly speaking we live in an organized, clockwork universe governed by precise mathematical equations, the end result is chaotic, and for all intents and purposes impossible to predict—or therefore to control.

For a first-cut look at streamflow memory, though, we'll use a different iterated map: $Q_t = \alpha \, Q_{t-1} + \varepsilon_t$. Let's walk through the parts of this simple algebraic equation. The symbols for streamflow and time are just as we described above, so $Q_t$ is today's streamflow, and $Q_{t-1}$ is yesterday's streamflow. The Greek lowercase letter "alpha," $\alpha$, is a fixed model parameter or constant, loosely like $r$ in the logistic map. It's usually called the serial correlation coefficient, and it multiplies yesterday's streamflow. The role of $\alpha$ is to determine how powerful the memory is—by adjusting its value, we can tune the model so that the system has either a strong memory (larger $\alpha$) or a weaker memory (lower $\alpha$) of what came before. We'll explain how that works in a moment. Finally, the Greek lowercase letter "eta", $\varepsilon$, represents a noise term, and is itself a time series too. By noise, we mean a random number (or at least something that we can treat as a random number, and get away with it most of the time). Each new day ($t$) gets a new random number ($\varepsilon_t$). Each of these daily random numbers has nothing to do with the previous day's one. The only thing they have in common is that it's usually assumed that each is drawn from the same distribution (this is called the stationarity assumption), which means that each possible value has a particular, and fixed, probability of happening—kind of like how the odds of drawing the ace of spades from a full and fair deck is always 1 in 52. This type of noise is called white noise, and it's exactly what it sounds like: basically, static. Genuine white noise is inherently completely unpredictable. So, if we set the constant, $\alpha$, to 0, then the equation simply becomes $Q_t = \varepsilon_t$, and the streamflow time series would just be totally unpredictable white noise, with no memory at all.

If instead we set $\alpha$ to 1, then we get $Q_t = Q_{t-1} + \varepsilon_t$. That's called a random walk. Probably the most widely known application of the random walk model is for daily stock prices. Although it's a strongly incomplete description of the stock market, and placing excessive confidence in it stands to get you in a heap of trouble, it's productively used in a variety of things, such as setting prices for derivatives. Maybe you've heard of the famous, or perhaps infamous, Black-Scholes options pricing model for hedging; it assumes that the stock market follows a particular type of random walk. The basic gist of a random walk is as follows. Imagine you're on a staircase. You randomly decide to take a step up or down, carrying you to a new step. Then you randomly decide again to take a step up or down, carrying you back either to your original step or to another step farther up or down from it. This process just keeps on going. Now replace yourself in this mental picture with the daily value of, say, the S&P 500 index, and you get the general idea. An important aspect of a random walk is its dual nature. There's the random jump at each time step. But your current position is also strongly determined by what your last position was. In this sense, a random walk has very strong memory. It remembers where it was last.

It turns out that the most usual case for daily streamflows, though, is to have a moderate amount of memory, represented by a value of α somewhere between 0 and 1. This kind of process gets another name: the first-order autoregressive process, or AR(1) process for short. These equations were first studied systematically by the late nineteenth- and early twentieth-century mathematician Andrey Markov, and they are often referred to alternatively as Markov processes in his honor. Like the random walk, an AR(1) process has a dual nature, containing both a totally random term (the $\varepsilon_t$ part), and a deterministic term (the α $Q_{t-1}$ part). It is the latter that gives the system its memory: the river's level today $(Q_t)$ depends partly on what it was yesterday $(Q_{t-1})$. Note that while the equation only explicitly considers yesterday's streamflow (hence the name, "first-order"), yesterday's flow in turn depended on flow the day before that, which depended on flow the day before that and so forth back in time. So the net memory often extends considerably more than one time step into the past. This important deterministic aspect notwithstanding, the influence of randomness on the outcome is such that an AR(1) time series is still commonly called a random process, and the end result is still called noise—but red, rather than white, noise. It may seem strange to assign colors to noise. In fact, the term "red noise" is a reference to the colors of the visible light spectrum; we'll get back to just what that means in a later chapter.

The AR(1) model has long been used as a streamflow forecasting tool. It accomplishes this task by taking advantage of memory. Specifically, if we switch around the subscripts a little, and take the average (usually written μ) of the noise component as its best-guess value, we then get the forecast equation, $Q_{t+1} = α Q_t + μ$, where $t$ is today and $t + 1$ is tomorrow. Much better and more sophisticated river forecast models are available now, of course, but virtually all of these still take advantage of memory in one way or another. And intuitively, such memory (also called persistence or inertia) seems reasonable: as noted earlier, most rivers don't switch from late-summer base flows to annual peak flood flows from one minute to the next. Persistence varies from one region, though, and even one watershed, to the next. Larger rivers tend to have more memory (larger $\alpha$) than smaller streams, for instance; on the other hand, desert rivers are prone to flash floods, with flows that can change very quickly (smaller $\alpha$). As an example, a typical $\alpha$ value for a medium-sized Pacific Northwest river might be around 0.7 or so.

The enquiring and critical reader may have noted by this point that much of the foregoing uses statistical ideas, and may reasonably question the results in light of Mark Twain's famous comment regarding lies and statistics. But in fact, there is a strong basis in physical reality for this approach. Memory and storage amount hydrologically to much the same thing. Such landscape features as lakes, wetlands, and deep porous soils and rock capture and temporarily store rainfall or snow-melt, and then slowly release it to the stream channel. So the immediate effects of storm events, for example, are attenuated: much of the rainwater goes into storage instead of directly into the stream channel. In a similar way, though, streamflows remember those storm events for a longer period of time than they would otherwise; afterward, the rainwater from a storm event slowly dribbles out of storage into the stream channel.

Let's examine this storage a little more closely. To recap, streamflow memory is the tendency of flow at one point in time to resemble flow at a closely preceding point in time. Most streams don't go from late summer low-flow conditions to over-bank flood conditions and back again in an hour—changes in flow take a bit of time, even if the weather changes very rapidly (as it might with the passage of a thunderstorm, for example). The reason for this is watershed storage. It's sort of like a bank

3.5. Wetlands come in many shapes and sizes. The Pantanal is the world's largest tropical wetland, with an area of about 140,000 km² covering parts of Brazil, Bolivia, and Paraguay. Its waters flow through several rivers to reach the Atlantic near Buenos Aires. Photo: Alicia Yo / Wikimedia Commons.

account. Some money goes in when it's payday, and some comes out when you pay a bill—but the total sum in the account varies relatively smoothly, insofar as it generally doesn't go from zero to flush and back again from one day to the next. This is because there's (hopefully) a reasonably substantial amount of money stored in the "reservoir" of your bank account that pads the day-to-day variations. Things like aquifers, lakes, and wetlands play a similar reservoir-like role for watersheds, moderating the total upward and downward variations in flow by absorbing incoming moisture, storing it, and releasing it when necessary. It's also the reason why most rivers don't go dry after it hasn't rained for several weeks. These storage mechanisms also perform other important environmental functions. Wetlands are a great example; they filter contaminants from the water and serve as critically important wildlife habitat.

And here's an important take-home message about time series memory and streamflow. It's been shown that by replacing or covering up such natural, storage-inducing landscape features with impervious surfaces like parking lots and rooftops, watershed urbanization can amount to a form of environmental brain damage, whereby the river's short-term memory is partly lost—expressed as a decline in the serial correlation coefficient we mentioned above. The asphalt and concrete and roofing tiles prevent rainfall from accessing the soil, and often, storm sewer systems convey runoff directly to the stream channel. As a result, natural water storage features like lakes, wetlands, soils, and aquifers receive a smaller proportion of the falling precipitation. With a smaller amount of accessible storage, $\alpha$ decreases, and the river can't remember quite so well what the previous day's flows were. That translates into increases in peak flow (along with things like flooding and erosion) during storm events, while also decreasing base flows (along with things like day-to-day availability of water supply and fish habitat) during interludes in the weather. It can also create more seasonal-scale hydrological shifts, resulting in lower water levels during what are already natural low-flow periods, such as late summer in much of the western United States. Overall, the river gets flashier and more unpredictable, with higher highs and lower lows. That's bad for both people and ecosystems.

But the broader plot thickens. Although linear approximations can often prove serviceable in practical applications, most systems—physical, biological, and social—are fundamentally nonlinear. What does that mean? The AR(1) model we've described above is linear. A linear system is one in which the output is proportional to the input. Consider a trip to the local grocer. There's a linear relationship between the total amount of apples you buy and the total amount of money you spend on those apples. If you plotted how much money you spent against how many pounds of apples you received, you'd get a straight line (hence "linear"). The sensitivity of apples bought to money spent would be the slope of that line, expressed in terms of the unit price of apples—say, a dollar a pound. What's more, that sensitivity wouldn't change depending on how many apples you bought or how much money you spent: regardless of whether it was two apples or twenty, your bill would simply go up in proportion. In contrast, a nonlinear system is one in which the response is out of proportion to the stimulus. To extend our apple example, consider the economies of scale you might enjoy by buying larger amounts in bulk. A graph of how much you spent against the total weight of apples bought wouldn't be a straight line anymore (hence "nonlinear"), but curved instead. That would mean the price per pound—the sensitivity of apple expenditure to total amount bought—would be different depending on whether you bought twenty apples, or two thousand apples. Important real-life examples of nonlinear effects are many. One is the exponential growth over time in population size. Another is the so-called tipping-point effect, whereby small incremental changes in forcing (say, habitat loss) have only a modest ecological impact, until the accumulated small changes eventually reach a critical level at which the system (say, fish population) suddenly and unexpectedly collapses under just one additional, seemingly minor instance of habitat loss.

The difference between linearity and nonlinearity is one of the most fundamental and widely important ideas in all of science, and especially in physics and its applications to the environmental sciences, so let's take a little bit of a closer look. Consider the graphs in figure 3.6. These are just schematic pictures show-ing the overall shapes of some possible examples of relation-ships between two variables, $x$ and $y$. These variables could be whatever you like—engine size and horsepower, perhaps, or

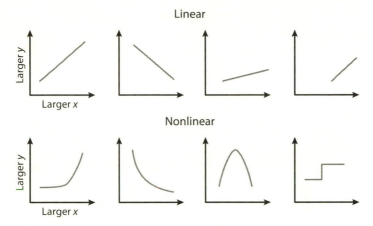

3.6. Examples of linear and nonlinear relationships between two variables, $x$ and $y$.

glasses of wine and stress level. Linear relationships can be increasing or decreasing, have different slopes, or get shifted around the graph. But as the name implies, they always consist of a single straight line. Nonlinear functions, in contrast, involve lines that change shape—they may be curved, or have sharp bends, for example—so that the basic way $y$ varies with $x$ depends on the size of $x$. For instance, in the left-most examples of both linear and nonlinear relationships in figure 3.6, $y$ increases with $x$. This increase is steady for the linear case. In the nonlinear case, $y$ starts off increasing very slowly with $x$, but then really takes off at large values of $x$. Exponential population growth and compound interest are examples of nonlinear relationships like this. We'll also note that if the functional form is simply a single flat-lying line (not shown), that means there is, in fact, no relationship between $x$ and $y$. That may not sound especially exciting, but it can actually be a valuable scientific result, as we'll see shortly.

What do linearity and nonlinearity mean for the way that rivers remember? For our AR(1) streamflow model, the assumption of linear memory means that today's flows are in proportion to yesterday's flows (at least as far as the nonrandom part of the equation goes). Nonlinear memory implies that the degree to which a river remembers its past flows depends on what ballpark the flows were in recently. To get a better feel for how this works, we'll consider two rivers, one in the United Kingdom and one in Canada. First, though, we need to find a way to identify and describe this nonlinearity.

And that's where artificial intelligence comes in. A variety of AI technologies have been developed, and the most broadly used is the artificial neural network, sometimes abbreviated ANN. There are in turn different kinds of neural networks. The most heavily used of these is called the multilayer perceptron, which has a wonderfully campy 1960s science fiction movie ring to it. The overall idea behind a neural network is to loosely emulate the function of the human nervous system. The earliest versions were often implemented in electronic hardware, whereas modern ones are generally in software—that is, computer programs.

Computationally, artificial neural networks are built in such a way that a lot of things can go on at the same time, and each part of it has something to do with all the other parts; this is called a fully interconnected parallel architecture. A schematic drawing of a typical multilayer perceptron, usually abbreviated as MLP, is shown in figure 3.7. MLPs are usually split into three layers: an input layer, into which data are fed; an output layer that gives you your answer; and a hidden layer that does much of the computational heavy lifting. Each layer consists of a number of neurons, named after the structures in the human brain that inspired this type of AI technology, and all the neurons in a given layer are connected to all the neurons in the next layer, somewhat like the neurons in our brains are connected by synapses. In an artificial neural network, these connections are made by means of a bunch of equations that involve, among other things, something called an activation

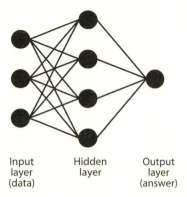

Input          Hidden          Output
layer          layer           layer
(data)                         (answer)

3.7. Schematic diagram of a feed-forward neural network, a type of artificial intelligence technology frequently used to model geophysical and other systems.

function. The activation function is a very rough mathematical representation of the process of a human neuron firing. Modern MLPs use activation functions that are nonlinear, which is a big part of the reason why neural networks are so powerful. In fact, if an MLP is appropriately configured, and properly "trained" (as with human brains, this is often the toughest part!), it's in principle a universal approximator that can model anything—at least, anything capable of being modeled deterministically. MLPs and other closely related artificial intelligence technologies are widely viewed as the ultimate black-box prediction tool, and because of those nonlinear activation functions, their capabilities include capturing and modeling nonlinear relationships.

That generality is important, because it allows us to apply ANNs to our question of streamflow memory. In forecast mode, the input layer consists of a single neuron, containing today's streamflow, and the output layer also consists of a single neuron, containing the prediction for tomorrow's streamflow. But all these neurons don't just automatically know what to do. As we alluded to above, just like the biological "neural nets" that you and I have tucked safely away beneath our skulls, artificial neural nets have to be trained. For MLPs, this is done using a so-called supervised learning algorithm, which involves using historical streamflow data to configure the network. Associated choices include picking the best number of neurons to have in the middle layer and finding the best-fit values for various parameters within it. It's an iterative process: lots of different combinations are tried, and the ones that do the best job of minimizing the error in the predictions are used to guide subsequent choices toward an optimal end point. And its duration is governed by the principle of diminishing returns: the learning is terminated when the improvements from one trial to the next become too small to be worth the bother. The result is then tested using a separate subset of the historical data, to help ensure that it really is capturing nonlinear streamflow memory and correctly using it to forecast tomorrow's flows. In practice, more effective methods are used to forecast streamflow, including but not limited to much more complicated ANNs. We'll return to that in some detail in a later chapter. Still, this simple ANN forms one of the basic building blocks often used in sophisticated operational forecasting systems. By the same token, this particular type of neural network yields some interesting physical insights.

ANN-predicted flow time series, and the relationships between flows one day ($Q_t$) and the next ($Q_{t+1}$) as determined by a neural network, are shown in figure 3.8 for two salmon streams on opposite sides of the world. One is the River Wye, which originates in the Cambrian Mountains of Wales, flows through grassland, moors, and forest, and eventually forms the border between Wales and England at its downstream reaches before reaching the Atlantic Ocean. Its headwaters lie near those of the River Severn, the longest river in the United Kingdom. The other is the Koksilah River, located on southeast Vancouver Island, a large island occupying much of the Pacific coast of British Columbia. The watershed consists mostly of temperate rain forest, rolling hills and mountains, small farms, and a few

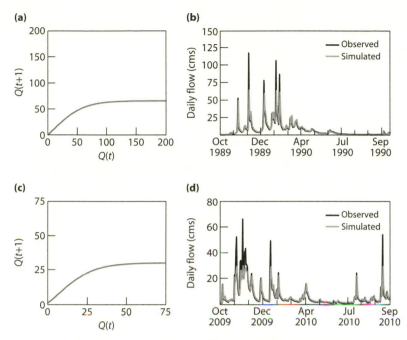

3.8. Nonlinear memory in the Koksilah River (top row) and River Wye (bottom). (a) and (c) The relationship between observed flow today, $Q_t$, and neural-network-predicted flow tomorrow, $Q_{t+1}$, isn't a straight line—the memory is nonlinear. (b) and (d) Simple one-day-ahead river flow forecasts based on nonlinear memory, and what actually happened the next day; results are shown for one year.

small towns. The lower reaches of the Koksilah River and the adjacent Cowichan River share a floodplain and delta, which happens to be, in my biased opinion, one of the most tranquilly beautiful places around; the delta then flows into the Pacific Ocean by way of the Salish Sea. Both the Wye and the Koksilah were once notable salmon fishing waters, and both have declined in that regard. Salmon have complicated life histories, spending much of their lives in the ocean but returning to freshwater to reproduce, which makes it particularly difficult to disentangle various different controls on the health of their populations. That said, freshwater habitat quantity and quality are known to be key influences, so understanding the hydrological dynamics of their rivers can be important—including, potentially, the nonlinear memory characteristics of river flows.

If these rivers had linear memories, the relationships between $Q_t$ and $Q_{t+1}$ on this plot would be a straight line. As you can see in figure 3.8, they're not. For both rivers, the graph is curved instead. The memories of the Koksilah River and the River Wye are, evidently, nonlinear. On the one hand, when flows are generally low, both rivers seem to follow almost a random walk, with strong memory. This corresponds to the steeply rising, nearly straight line at the left-hand sides of the graphs for both rivers. But when flows are generally high, the river instead adopts a more or less white noise pattern with very little memory. This corresponds to the approximately flat line at the right-hand side of both graphs. This doesn't mean that $Q_{t+1}$ is nearly constant at high $Q_t$, but instead that the relationship between the two largely disappears. That is, at high flow, river discharge fluctuations can temporarily become quite random, indicating that $Q_{t+1}$ has nothing to do with $Q_t$, and would instead be driven by $\varepsilon_t$, the white noise component of the AR(1) process.

Why does this nonlinearity occur, with strong memory (in fact, almost a random walk) at low flows, and weak memory (in fact, almost white noise) at high flows? Physically, it has to do with the fact that different streamflow-generating mechanisms kick in under different conditions. When flows are low, much of the water present in the river channel is generated by release from the storage mechanisms we mentioned earlier—landscape features like ponds, wetlands, lakes, and shallow aquifers, which are a comparatively nice, steady source of water. In contrast, high flows tend to occur under more extreme weather conditions, like storms. Heavy or sustained rain occupies all the available storage, so that any additional precipitation landing in the watershed has nowhere to go, except to scoot quickly to the stream channel. The river becomes, in effect, more directly coupled to the weather, which of course is relatively changeable and tougher to predict—hence the poorer forecast skill at high flows using a memory-based model (see figure 3.8). That is, low flows are dominated by watershed storage and therefore exhibit strong memory, whereas at high flow, river discharge fluctuations can temporarily become seemingly random, as watershed storage has been overwhelmed and chaotic weather processes drive the flows.

It's thought that this sort of nonlinear memory in river flows is widespread. Just how widespread, though, remains an open question—as do the details of its origins, its relationships to watershed storage factors like basin size and geology, or how it might be influenced by urbanization and other land use changes. Nonlinear memory is an important mathematical manifestation of the broader hydrologic fact that a given river can react to a given weather system very differently depending on how wet or dry it's been in the hours, days, and weeks leading up to that storm. Getting a better grasp on these issues is fundamental to our understanding of the time series properties of river flow, with implications that could range from how to improve the accuracy of flood forecast systems, to more thorough evaluations of fish habitat quantity and quality in salmon rivers like the Koksilah and Wye.

That is, there is still a great deal to be learned. This is typical. Almost every scientific step forward raises at least as many questions as it answers, bringing us to a fundamental tension at the core of scientific discovery. On the one hand, the main goal of science is to remove mystery by distilling the physical world into a concise, often mathematical, blueprint, and it's very good at its job. There are many kinds of truth, and many legitimate paths to those truths. The evidence-based and self-correcting systems of scientific pursuit, however, are such that perhaps no other mode of human inquiry specifically reveals the hidden physical mechanisms of nature in such a consistent, precise, objective, and generally effective way. On the other hand—and scientists tend to speak less of this—science requires and even rejoices in mystery. With no mysteries to solve, there is no science to be done. More than that, though, science is an expression of our wonder and curiosity about the unknown. It is a human pursuit, and as such it does not generate a monument of final and absolute truth, but something a little more organic and changeable and responsive to will and whim. Mystery is what gets scientists out of their beds and to work in the morning. The choices of which mysteries to solve and how to solve them are often subjective and individual, and the history of science charts out a winding and unpredictable path deeply influenced by personal inspiration. But the juxtaposition between these two responses to the unknown—to eliminate it, and to celebrate it—is easily accommodated: there is infinitely more that is unknown than is known. Science is the art of proximal answers. It reveals the cause of some effect, but that cause is in turn the effect of some other cause yet to be discovered, and so forth. The already astonishingly large, detailed, and rigorous body of scientific knowledge will continue to grow, yet we will, thankfully, never run out of mystery.

# 4

## CLOUDS TALKING TO FISH:
## THE INFORMATION CONTENT OF RAIN

Every day, the news announces another invention or implication of the information age. Some of these changes are good, and some are not. Certainly, the world is a much smaller and less private place than it was only a decade or two ago. But how has this digital age transformed environmental science? Some of the consequences are relatively obvious: just like consumer electronics, environmental monitoring technologies are benefiting from improved capabilities and lower costs. Similarly, the availability of increasingly brawny yet inexpensive computational power has widely enabled types of data analysis and modeling that previously were prohibitively expensive or simply impossible. However, there is also a subtler, and perhaps more profound, way in which the information age is affecting environmental science. It offers an entirely new way of viewing the natural environment—as a vast and complex information processing system, which can be described using the same mathematical theories that underpin the technologies of the information age. This view has the potential to substantively change our conceptualization of the natural environment, enabling us to see problems differently or even to recognize new problems entirely. Let's explore how this could work by taking a look at what these ideas, transplanted from communications theory, might be able to tell us about the way that weather, water, and fish interact.

Clearly, there are close relationships between fish and river flows, though the nature of these interactions varies widely. It's worthwhile to take a moment to consider three examples: a tropical jungle, a northern rain forest, and an arid desert. The Orinoco River of Colombia and Venezuela lies just to the north of the Amazon. It's one of the largest river systems in the world, spanning more than 2,000 kilometers in length before it empties into the Atlantic Ocean just east of the Caribbean. The Orinoco experiences a strong yearly cycle with high flows from April to October, corresponding to a tropical rainy season. Piranha are notoriously voracious predatory fish. But in this basin, one member of the piranha family synchronizes its reproduction with the arrival of the high-flow season to take advantage of a vegetarian food source. Rising water levels flood the forests adjacent to the river, making them available to the piranha, which then feed on fruits and seeds. While the vegetarian piranhas of the Orinoco may be an oddity, hydrologic seasonality like this has deep ramifications for virtually any aquatic ecosystem worldwide.

Salmon runs are another great example of water-fish relationships. Salmon are, in general, anadromous: that is, they spend much of their lives at sea, but they're born in freshwater, and at the ends of their lives they return to the stream of their birth to reproduce in turn. Low flows during the spawning migration can compromise reproductive success by drying up the "fish highway" from the sea to the spawning grounds. Lower river flows also tend to mean warmer water, which can also have implications for reproductive success because salmon are cool-water species. Conversely, high flows can excavate and destroy shallowly buried salmon eggs during their incubation period. This in turn affects not only the aquatic ecosystem but also a web of interrelationships extending far past the water's edge. For example, along the much of the Pacific coast of North America, bears, wolves, and eagles not only feed on spawning salmon but also drag the carcasses into the adjacent woods, where their decomposition fertilizes the forest.

But perhaps nowhere else is the influence of the freshwater cycle on life as striking as it is in the desert, where the simple presence or absence of water can have a dramatic impact. Consider the arid regions of the American Southwest. The greater soil moisture beneath dry washes, the intermittently flowing creeks that dominate desert stream networks, support phreatophytes—plants, like the desert willow, that send their roots deep to capture groundwater—controlling the spatial distribution of vegetation across the landscape. And a particular variety of shrimp depends on "tinajas," pools of water that form following a rainstorm. The so-called tadpole shrimp can survive literally decades of drought as an egg, and then hatch and reproduce in as little as two weeks when the hydrological conditions are right.

The study of hydrology-ecology relationships like these is often called, reasonably enough, hydroecology, though sometimes it may go by other names as well. Whatever you choose to call it, the field typically brings several techniques to bear. For example, statistical methods are often used to identify the complex web of cause-and-effect relationships. Process-based mathematical models, which attempt to simulate habitat and hydroecological system behavior by a bottom-up approach explicitly representing what is known about all the individual mechanisms involved, can also be used. And particularly in remote environments, where aboriginal peoples such as the Inuit still maintain intimate ties to the land via subsistence hunting and fishing, studies of traditional ecological knowledge have also proven fruitful, albeit sometimes challenging to integrate with the usual methods of science.

But let's explore a different way of looking at how weather, water, and ecosystems are interrelated. Although many other scenarios could be entertained, consider for relative simplicity a small watershed driven mainly by rain events. So what we have here is rainfall helping to drive river flows, and river flows in turn helping to drive fish populations. Or perhaps more to the point, fluctuations in rainfall drive fluctuations in river flows, and those fluctuations in river flows help drive fluctuations in fish populations. Each of these things—rainfall, streamflow, and fish population—can be quantitatively measured, forming data sets. Data contain information about the thing being measured—that's why we bother to collect them. So our conceptual model involves, in effect, the transfer of information from the atmosphere, to the watershed, to the fish species present within it. And the transfer of information, of course, is the essence of communication.

That is, clouds talk to fish, using the watershed as a communication channel like a telegraph wire or a fiber optic cable or a cell phone transmission. This way of perceiving environmental relationships isn't just some fanciful act of anthropomorphization. We can back it up with math—specifically, with information theory. The foundations of modern information (or "communication") theory were laid down in a 1948 paper published by Claude Shannon in the *Bell System Technical Journal*. Shannon was a mathematician at Bell Labs, working on problems of telegraphy and telephony, critically important practical issues of the day. Not only were the principles set out by Shannon and his predecessors sufficiently general to survive the transition to the digital age, they also were in fact superbly suited to digital communications and computation, and indeed played an important role in ushering in that new age. Much of the technology surrounding us today involves, directly or indirectly, Shannon's theory of communication. At the heart of that theory lies something called the Shannon entropy, to which we'll devote much of this chapter. It's a measure of information content. And while the difficulty and value of being able to measure the quantity of information using a simple formula may not be immediately apparent, it turns out that inventing this measure involved a remarkable conceptual trick and had tremendous ramifications. Indeed, one of the most amazing things about this information theoretic entropy is the extraordinarily wide range of fields that it has diffused into—many of which, at first blush, might seem to have very little to do with telecommunications. Linguistics was one of the first, and indeed Shannon's original paper hinted at those possibilities. Many others have since emerged, some of which we'll talk about below. It's yet another example of the complex conceptual interactions between seemingly unrelated sciences that can help shed new light on old subjects.

Another field where Shannon entropy has had a great impact is in physics. There are likely several reasons why physicists have latched on to the concept. One is that there is overlap between the "entropy" of information theory and the "entropy" of thermodynamics and statistical mechanics, a key idea in physics. Another is that physicists have grown steadily more interested over the last two or three decades in the analysis and prediction of complex systems. As it turns out, Shannon entropy and other similar quantities can be used to gauge a system's complexity and predictability, with applications ranging as far afield as quantitative finance and computational sociology. One of the basic notions here is that the complexity of a system can be measured by the amount of information required to encode its behavior (or in a conceptually related measure called the Kolmogorov complexity, the length of the optimal algorithm or computer program required to recreate its behavior). But perhaps the most compelling reason for the penetration by Shannon entropy into mainstream physics, and other fields as well for that matter, is also the subtlest. In his wonderful popular science book, *The Bit and the Pendulum*, Tom Siegfried explores the idea that a society's dominant new technologies or machines at any given time, be it clocks or steam engines or computers, provide metaphors that can inspire fresh and productive ways of asking and answering questions in science:

> Today the notion of a "clockwork" universe is synonymous with Newton's physics. . . . Newton's science did not create the clockwork metaphor, but exploited it . . . since its invention half a century ago, the electronic computer has gradually established itself as the dominant machine of modern society . . . the defining feature of computing is the processing of information, and in research fields as diverse as astrophysics and molecular biology, scientists . . . have begun using the metaphor of information processing to understand how the world works in a new way.

That is, we live in the so-called information age, and the notion of viewing the world through the lens of information (and its mathematical accoutrements, like Shannon's theory of communication) has spread to scientific research on subjects having little if anything to do with telecommunications. That is, how science is done can depend heavily on how you conceptualize whatever it is you're studying, and that is in turn heavily influenced by both the technologies and the culture of the day. In Newton's time, the idea of a clockwork universe was pervasive, partly for philosophical reasons, and it seemed natural for him to tackle problems like gravitation by phrasing them in that way. Today, information—the hard reality, in terms of communications and big data for instance, as well as the basic notion of looking at everything as a source or store of information—is the zeitgeist, and the phrasing of scientific questions has adjusted course accordingly.

Some of the corresponding applications have been esoteric, to say the least. Consider the tongue-in-cheek British book and television series called *The Hitchhiker's Guide to the Galaxy*. One of its storylines involved an advanced race of alien beings who built a sentient supercomputer that, it was hoped, would deduce the meaning of life. After a very long time, it came back with the answer: 42. When the puzzled aliens pressed the computer for an explanation, it told them the answer was right, but they had asked the wrong question. In an effort to determine what the right question was, they proceeded to build an even grander computer: Earth. (Unfortunately, just as it was about to yield the solution, our planet was demolished by a different, more modestly advanced race of alien beings to make room for an interstellar superhighway.) Certain interpretations of nature, constructed by physicists using the principles of information theory, may seem rather similar to this whimsical sci-fi view of Earth as a supercomputer built to figure out "the right question," but they are in fact quite serious. These include mathematically formal views of black holes (and in fact the universe itself!) as constituting information processing devices—in a word, computers.

More down-to-earth applications of Shannon entropy can be found in hydroecology, or at least in its constituent disciplines. For example, even some introductory-level textbooks in freshwater ecology include the so-called Shannon-Weaver biodiversity index—although its information theoretic origins may not be explained, and one could be forgiven for concluding that Claude Shannon was a biologist! (Weaver, by the way, wrote an excellent and widely read introduction to a book that Shannon published shortly after his key paper.) The notion in this case is that a healthy, complex ecosystem having a rich abundance of each of many different species contains a great deal of biological information. At the opposite extreme, a simple or impoverished ecosystem dominated by just a couple of species has relatively little biological information content. Intuitively, this makes sense. Look at it in terms of the number of Latin species names you'd have to memorize for a test—more studying would be required to answer an exam question on the richer ecosystem because there are more species present. Another relatively common environmental application of Shannon entropy resides with the optimal design of observational networks, such as regional networks of rain gauges, stream gauges, or groundwater monitoring wells. The corresponding interpretation of information content is quite literal. As we noted earlier, the idea behind gathering data is to obtain information. Shannon entropy is a measure of the information content of a data set. So it's perfectly reasonable to use entropy as a yardstick when selecting the set of particular locations where a given number of available rain gauges, say, should be placed such that they yield the greatest amount of information.

We should take a moment here to emphasize that the establishment, optimization, and long-term operation of such observation networks is absolutely key to understanding rivers and the environment more generally. Data are the basic fodder of all science, including hydrology. As alluded to above, one important application of information theory is in hydrometric network design, which involves deciding where to best place the gauges used to measure river flows. These gauges use various technologies, but most commonly, the water level of the river, known as its stage, is measured using a pressure transducer that essentially weighs the column of water above the streambed. Stage is then converted to a streamflow measurement using mathematical relationships called rating curves, the initial development of which requires detailed measurements of water velocity for a given gauge location. The consequences of not collecting a sufficient amount of such data—in terms of both the total number of gauges and how long they're operated—can be painful and long-lived. Consider the Colorado River Compact. This agreement, signed in 1922, apportioned the Colorado River's flow between seven US states. Allocations

**4.1.** Arizona's Lake Mead, formed by Hoover Dam on the Colorado River. The top of the white band is where the reservoir rose to when water was more abundant; record low levels were experienced in 2015 when this picture was taken. Photo: S. W. Fleming.

were based on the short environmental data sets available at the time. It turns out that the period covered by those records sampled a climate phase called the North American Pluvial, revealed by subsequent paleoclimate reconstructions (using carefully measured variations in the widths of tree rings) to be one of the wettest periods in the region in centuries. The resulting overestimation of available water led to overallocation and overextraction, and in most years, the river no longer flows to its mouth over the border in Mexico. Using information theory or other quantitative tools to optimize the design of these environmental monitoring networks is one of the most common, yet most important, tasks that hydrologists undertake.

On the other hand, it's harder to come by examples of the direct application of communication theory to examination of broader hydroecological relationships. Below, we'll limit ourselves to considering just one component of that larger web of hydroecological interactions. Specifically, after explaining how Shannon entropy works, we'll show how it can be used to quantify the information content of rain, explore what that might hint at in terms of atmospheric information-generating processes, and then return to what the outcomes might imply for the relationships between weather, water, and fish.

The essence of Shannon entropy lies with the equivalence of information and uncertainty. At first glance, this idea will likely seem profoundly counterintuitive. After all, aren't information and uncertainty complete opposites? The answer is that it's a yin-and-yang thing: just as light cannot exist without dark, there can be no information without uncertainty. Information and uncertainty are indeed diametrically opposed, but you can never gain any useful new information if there isn't some kind of uncertainty to be resolved. Put another way, getting your question answered doesn't give you any new information if you already know what the answer is; if you already knew the answer, there was no point in asking the question! And the amount of information you obtain by monitoring a signal can therefore be expressed in terms of how much uncertainty was removed by doing so. That doesn't mean that information and uncertainty are fundamentally the same thing, because obviously they aren't, but Shannon gained traction on the problem by making that simplifying assumption and then running with it.

Let's take a concrete example. Say you make daily measurements of rainfall at some particular location. One day you might get 0 mm, the next day you might get 37 mm, the day after that might bring a total of 5 mm of rain, and so forth. As we've discussed in a previous chapter, such a series of observations of something over time is called a time series, and the analysis and prediction of such data sets is called—reasonably enough—time series analysis and prediction. Now, let's divvy up those daily total rainfall measurements into, say, four bins: none (a dry day), a light-rainfall day, medium rainfall, and raining cats and dogs. This process of lumping numerical measurements into bins is variously called discretization, quantization, or in the nomenclature of symbol sequence analysis, coarse-graining the data into an alphabet. All these colorful terms mean essentially the same thing. Here, we do it mainly for simplicity, but it's worth noting as an aside that there can also be more fundamental reasons for discretizing time series values. Anyway, let's denote each of the above four rainfall states by its abbreviation, that is, by a symbol: N (none), L (light), M (medium), and H (heavy). The resulting symbolized daily rainfall data set over, say, eighteen consecutive days might look something like NHMHNNLLLLNNNNNLMH, for instance.

Now, at one extreme consider some imaginary place where, every single day, the rainfall comes in as "medium." The resulting symbolized time series is nothing more than a long string of Ms. In this simple hypothetical case, daily rainfall is, always has been, and presumably always will be "medium." How much uncertainty is there in daily rainfall amounts here? None: it's always "medium." And how much information do we gain by hanging around to measure the rainfall tomorrow, and the next day, and the day after that? None: we already know what the answer is—"medium." Information content is the degree to which uncertainty is removed by listening to or monitoring a message or signal, and if there is no uncertainty to be removed, then no information has been communicated or gained. Now consider the opposite extreme, in which the four states—N, L, M, and H—are all possible and in fact all equally likely. In this case, you have no idea what the next measurement will be—it could just as well be any of the four possibilities—until you actually take it. There is uncertainty, and by hanging around to collect data, you remove that uncertainty. Information is obtained. These two examples are the simplest cases of minimum (specifically, zero) and maximum potential information. There are also intermediate cases, and although more complicated, they are also generally much more realistic, and we'll get to them shortly.

So now we have the essence of Shannon entropy: the equivalence of uncertainty and information. The basis of how entropy actually works in detail lies with the minimum average number of binary questions required to determine the current state. This is easier than it might sound. A binary question is a question having only two answers: yes or no, right or left, 0 or 1. Consider first the maximum-information scenario we entertained above, in which the answer is equally likely to be a daily rainfall amount of N, L, M, or H. We can use a decision tree to assess how many binary questions are needed to determine or "encode" which of these four possibilities was actually realized for a given day:

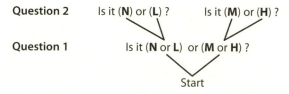

4.2. Binary decision tree for 4-bin, maximum-information case. Here, the result is that rainfall at this hypothetical location has an information content of 2 bits per measurement, or equivalently—because we're talking about measurements of daily rainfall totals—2 bits per day.

How do we use this decision tree? Say your buddies went out and measured today's rainfall as a favor for you. But now they've had a couple beers, they're feeling like giving you a hard time, and they expect you to play a guessing game to figure out the answer. Here's the nature of the game: you can only ask binary, yes/no-type questions. Assume for now that the possibilities for daily rainfall amount are as stated above: none (N), low (L), medium (M), or high (H), and that each of these is equally likely. What is the minimum number of such questions you'd have to ask your friends to determine which of these four bins today's rainfall belongs in? This decision tree shows that the answer is 2. For instance, say the measurement happened to be L. In this strange but ultimately very useful framework of asking binary questions, we'd first have to ask, "Is it N or L, or is it instead M or H?" Having determined that it's N or L, the next question would be, "Is it N, or is it L?" And then we'd have our answer: L. So that's the maximum-information case—and what about the zero-information case above? No binary questions are required, because the answer is the same every time (M, in the example we discussed a little earlier). Again, you don't have to ask the question if you already know the answer.

The answer to a binary question can always be represented as a 0 or a 1, and the answers to a series of such questions can be represented as a string of 0s and 1s. For the rainfall amount of L in the foregoing maximum-information scenario, when working our way up the decision tree shown in the corresponding figure above, the paths taken were "left" then "right" to arrive at L. So if we take "left" and call it "0," and we take "right" and call it "1," the answer would be "left-right" or 01. The number 01 is called a binary number, and it is in this case made up of a string of two binary digits, corresponding to the answers to our two questions. A binary digit is simply a numerical digit that can only take on two values, 0 or 1, and is often contracted to form the word "bit." Now, with the introduction of this simple piece of nomenclature, you might get a glimpse of how Shannon entropy has played such an important role in the digital age. Eight bits make a byte, and of course bits and bytes—along with subsidiary units like gigabytes (GB) of storage, megabits per second (Mbps) Internet download rates, and so forth—form our basis for measuring the storage and transmission of digital information. And this simple appearance of the word "bit" in both contexts is no accident: digital logic, computation, and communication work by flicking electronic circuits on and off, corresponding to 0 and 1, very much like the answers to our binary rainfall questions above.

And so, getting back to our daily rainfall example, the fact that we needed two questions means that 2 bits of information were gained per measurement. And in the no-information case, where the answer was always M and no binary questions were necessary, the signal contains 0 bits of information per day.

We can use these ideas to determine the information content of any rainfall time series, but in many cases, doing so will require a little something extra. Say each of the four daily rainfall states—none, low, medium, and high—all occur, but not all with equal frequency. For instance, N might happen 85% of the time, and L, M, and H might each occur only 5% of the days. In other words, on any given day, there's an 85% chance that the daily precipitation is zero, a 5% probability that it's light, and so forth. How many bits of information about rainfall, per measurement, does monitoring rainfall at this location provide? Well, how many binary questions do we need to ask to determine the rainfall state on any given day? The answer now is not quite so clear. On the one hand, as in our maximum-information case, all four states are very real possibilities. So it would seem we'd need two binary questions, and each measurement of daily rainfall would give two bits of information. On the other hand, if we simply assumed that the daily rainfall amount is always N without actually bothering to measure it, we'd be right 85% of the time—not too bad. And that would be an approximation of the zero-information scenario, in which we don't ask any binary questions at all, and receive zero bits of information. Clearly, then, this example lies somewhere in between the zero- and maximum-information extremes.

But where exactly? We need a more efficient, flexible, precise, and generalized way to answer it than drawing decision trees on paper. What we need, in short, is a formula. And that formula is Claude Shannon's groundbreaking achievement, an expression for information theoretic entropy, which gives us the minimum average number of binary questions needed to determine the system state:

$$E = -\sum_{i=1}^{S} P(x_i) log_2[P(x_i)]$$

We're not going to do any math here—if you're interested in the details of what this equation means, how to use it, and just where exactly it came from, there are plenty of useful reference materials listed in the "Further Reading" section at the back of this book I'd encourage you to consult—but we'll run through the basic idea. Shannon's equation for entropy takes our binary decision tree approach and codifies it in a way that's nice and flexible and robust. And the manner in which it does this is rather elegant. It just performs some relatively simple calculations on the probabilities of encountering each of the $S$ different possible system states (in our case, $S$ equals 4; the system states, $i$, correspond to N, L, M, and H; and in our last example their probabilities, $P_i$, are 0.85, 0.05, 0.05, and 0.05 respectively). The results are then added up (the uppercase Greek letter "sigma," $\Sigma$, usually denotes summation). That's all there is to it.

If we crunch the numbers through this equation for our last case, where all four rainfall states were possible but not equally likely, and we couldn't quite figure out how many binary questions we had to ask, we get $E = 0.85$ bits per measurement. That is, each daily measurement contains an average of about 0.85 bits of information about rainfall at the observation site. It might cause some consternation that the Shannon entropy can take on values that aren't whole numbers. How can you ask 0.85 binary questions? Either you ask a given question, or you don't. A way of looking at noninteger values of entropy lies with the idea of an *average* information content. Let's return briefly to our earlier, maximum-information example in which each of the four rainfall states is equally likely. But this time, let's start with defining not four, but six potential daily rainfall classes: none (N), very light (VL), light (L), medium (M), heavy (H), and very heavy (VH). The corresponding decision tree might look like the one in figure 4.3. In this case, the number of binary questions you have to ask in order determine the system state on a given day depends on what the state is. There's some arbitrariness in the way we drew the tree, but as given above, for N or M only two questions are required. Continuing with the convention of assigning a single binary digit for each choice, and setting it to 0 for a left branch taken and 1 for a right branch, we would encode N by the two-digit binary number 00 (left-then-left), and M by the two-digit binary number 10 (right-then-left). If the answer happens to be VL, L, H, or VH, though, we need to ask three binary questions. The encoding for H, say, would be the three-digit binary number 110 (right-right-left, going up through the tree). In a nutshell, then, the number of bits conveyed by each measurement, *averaged over all measurements* and therefore (in this case) all system states, would be somewhere between 2 and 3. So you can see that having noninteger values for $E$ isn't really a problem; we're still asking whole binary questions, it's just that the number of questions required varies a little, and taking the average of those whole numbers can give a noninteger value. In fact, using the formal expression for Shannon entropy, it's 2.6 for this particular problem.

By way of an addendum, if we compare this result to that obtained for four equally likely states (2 bits), we see that—all other things being the same—the more potential states into which we divvy up the signal, the greater the information content, even if the raw daily precipitation data are identical. This stands to reason. If we break down the original time series into finer segments, we're storing more information. If we know whether the precipitation is none, very low, low,

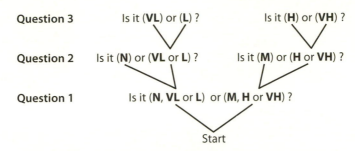

4.3. Binary decision tree for 6-bin, maximum-information case.

medium, high, or very high, then we know more fine detail about that daily precipitation amount than if we'd split it into the coarser intervals of none, low, medium, and high. (Note, though, that this doesn't necessarily imply that more and finer bins are always better.)

As with any scientific concept, there are plenty of caveats we could make regarding the meaning and use of Shannon entropy. There are also a number of other information measures available, some closely related, others not quite so much. We won't bother with most of those issues here, but there is one big point that must be made: Shannon entropy measures only the quantity, not the quality, of information. The significance of this fact should be apparent to anyone who's ever done an Internet search on some topic of interest, only finding a needle of good information (if that) buried in a haystack of useless or unreliable websites.

So far, so good. We now have in hand the fundamentals of Shannon entropy—a means for calculating information content, measured in terms of bits of information per measured data point, and one of the basic theoretical building blocks of the information age. But how can this concept speak to the weather-making atmospheric processes that drive rainfall, rivers, and freshwater ecology? What does it tell us about rivers? As we'll see, it provides a way to think about the range of mechanisms that generate rainfall and streamflow, and it may ultimately open up the idea of an optimal level of information content for fish.

Let's explore the question by considering daily precipitation data over a two-year period from two climate stations. One is situated near sea level in the Pacific Northwest (specifically, in the city of Vancouver), and the other is located in some nearby mountains (the Coast Range). The first order of business is to divvy up the data into bins. For simplicity, let's continue with the four rainfall classes we defined previously: none, low, medium, and high. The highest daily rainfall on record at these two stations over the period we'll consider here is about 133 mm, so if we split up the interval 0 mm to, say, 135 mm into three equal pieces plus a no-rainfall bin, we get the class intervals given in the table.

| Symbol | Description | Daily Rainfall Amount |
| --- | --- | --- |
| N | No rainfall | 0 mm |
| L | Light rainfall | Between 0 mm and 45 mm |
| M | Medium rainfall | Between 45 mm and 90 mm |
| H | Heavy rainfall | Between 90 mm and 135 mm |

We can then go through the entire rainfall time series for the low-elevation location, replacing the numerical rainfall value for each day with the corresponding symbol. For instance, daily precipitation amounts over the four-day period from September 27 to September 30, 2000, were 0, 6, 57.6, and 3.2 mm; using the above table, we'd replace that with N, L, M, L. Then we do the same thing for data measured at the high-elevation climate station, again using the same table.

So that's the discretization part. The next step is to find the relative frequency of occurrence, or probability, for each of N, L, M, and H for the low-elevation station. How do we do this?

Easy. Start by counting on how many days rainfall is "N," for instance, and dividing that by the number of days in the record we're considering, which is 731 (note that 2000 was a leap year with 366 days). The answer happens to be $P(N) = 0.4788$. That is, rainfall was "N" about 48% of the time over that two-year stretch. We repeat that procedure for L, M, and H. The whole process is then repeated independently for the high-elevation location. And yes—in case you were wondering at this point—as in many things worth doing, the practice of science is typically 1% inspiration and 99% grunt work. Results are shown below.

|        | Sea Level Station | Mountain Station |
|--------|-------------------|------------------|
| $P(N)$ | 0.4788            | 0.4679           |
| $P(L)$ | 0.5157            | 0.4802           |
| $P(M)$ | 0.0055            | 0.0451           |
| $P(H)$ | 0.0000            | 0.0068           |

From here on, the calculation is just plug and play; that is, take the probabilities from the above table, plug them into the equation for Shannon entropy we showed earlier, and work through the arithmetic to get an answer. And there we have it: Shannon entropy for daily total rainfall at the low-elevation coastal site turns out to be about 1.04 bits per day, but it's 1.27 bits per day in the nearby mountains. So high-elevation rainfall appears to carry slightly more information content than rainfall measurements at sea level in this case. Can we draw any broader conclusions from this observation? For example, might it tell us something about differences in weather complexity between low-elevation and high-elevation locations, and why those differences in information content exist? Mental exercises like this provide a great way to form hypotheses to subsequently test, and in our case, it also provides us with an opportunity to explore some important basic features of weather systems and rainfall patterns.

Let's start off by giving a proximal answer to why the Shannon entropy is higher for the high-elevation site. It lies with the evenness with which rainfall amounts are distributed across the four bins. Recall from our earlier examples that information is greatest when every category is equally likely, and least when only one class ever occurs. We can see from the list of probabilities in the table above that, although N and L are by far the most common rainfall amounts at both the high- and low-elevation sites, the frequencies of occurrence of different rainfall amounts are more evenly distributed across all four bins at the former than at the latter. Hence, there's a little more uncertainty about rainfall amounts in the mountains— consistent with the notion, familiar to any hiker, hunter, or backcountry skier, that mountain weather is unpredictable—so the signal measured there carries a little more information. In particular, the common categories of N and L are somewhat less common, and the rare categories of M and H are somewhat more common, at the high-altitude mountain site; it's just wetter overall, which in this case gives a more even distribution of rainfall frequencies across bins. But a deeper view of why Shannon entropy is higher at the high-elevation site may also be possible. Loosely speaking, rainfall at sea level here carries information generated by one very general type of weather process. In contrast, rainfall in the nearby mountains carries information about two.

Broadly, rainfall conditions over temperate areas of western North America are dominated by large weather systems that blow in off the Pacific Ocean. There are many types of large-scale weather systems. One involves the redirection of very wet, warm air from the vicinity of Hawaii, to form a long, narrow "firehose" of rainfall directed at the North American west coast. This is termed a Pineapple Express, and it commonly leads to rain-on-snow events, with a burst of both intense rainfall and snowmelt that gives the region some of its most severe local flooding events. One such Pineapple Express produced the so-called New Year's Day Flood of 1997, for example, which required the evacuation of more than 100,000 people and did more than $1 billion damage in northern California. The broader phrase, "atmospheric river," is used to describe phenomena like the Pineapple Express, which also occur in Europe and elsewhere. Conversely, sometimes cold, dry air moves south-westward from northern Canada in what are termed Arctic outflow events, bringing dry, clear skies and daytime highs well below freezing for a few days at a time to the Pacific Northwest in midwinter. But the most common scenario is a long series of low-pressure systems, born in the North Pacific, bringing an alternating sequence of moderate rainfall and short-lived dry spells. In combination, the foregoing types of weather patterns are largely responsible for the rainfall amounts experienced though much of the populated, lowland areas of the Pacific Northwest, including the low-elevation climate station considered in our analysis.

Exactly the same thing happens at higher-elevation locations. These sorts of storm systems and weather patterns are physically large—hundreds of kilometers across or, often, much more—and they don't magically stop just short of the mountains. But unlike coastal locations, high-elevation stations also experience a second kind of major control over the rain that falls there: orographic lift.

"Orographic" more or less means "having to do with mountains," and orographic lift refers to the fact that when a parcel of air collides with a mountain range, the air parcel always loses. Say air is blowing in on a westerly wind off the Pacific and hits a coastal mountain range. It then has nowhere to go but up. And when air is brought to a higher elevation, its pressure decreases. Think of it this way. When you're standing on some beach at sea level, all the air in the atmosphere above you presses down on you with all its weight. Of course, you don't actually notice

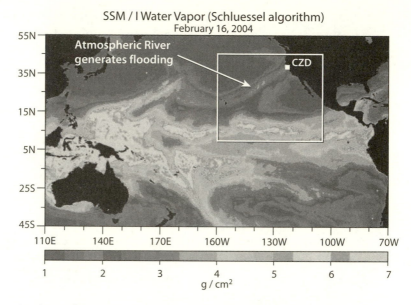

**4.4.** Atmospheric water vapor over the Pacific Ocean on February 16, 2004, with a Pineapple Express syphoning moisture from a wet band around the tropics and directing it toward central California. "CZD" is a weather station near San Francisco. Image: National Oceanic and Atmospheric Administration.

this, because people are built to live with standard atmospheric pressure, just like a whale is built to live with the otherwise bone-crushing water pressures deep in the ocean. Now go to the top of a mountain, a kilometer or two or three above sea level. You still have the weight of all the air in the atmosphere overhead pressing down on you, but there's less atmosphere above you now. So the total weight of that vertical column of air over your head is less—meaning that the air pressure is lower. And all other things being equal, when the pressure of a gas (like air) decreases, then its temperature decreases too. This follows from something called the ideal gas law, which you may or may not remember from high school chemistry, and it's called adiabatic cooling. Much the same thing happens when you pump up a flat bicycle tire, although it's in reverse so it's called adiabatic heating; the air you're forcing into the bike tire is getting compressed, and the air (and also tire, with which it's in contact) gets warm. In summary, then, an incoming air parcel gets forced upward by the terrain, it expands and loses pressure, and adiabatically cools. As it turns out, cool air can hold less water vapor than warm air, so as the air parcel rises and cools, water condenses out into liquid form and falls to the ground as rain. And this is why high-elevation spots generally are a lot wetter.

And with this we have a potential explanatory hypothesis for variations in information content across these different weather stations. Rainfall at the sea-level coastal location is controlled largely by large-scale weather systems. But rainfall at the high-elevation site, tucked away up in the nearby mountains, is controlled by both large weather systems and orographic uplift. Thus, rainfall measurements at the coastal site contain information about one very general type of atmospheric process, whereas rainfall at the mountain site contains information about two. And it would appear, at least on the basis of our cursory treatment, that this is duly reflected in the Shannon entropies for the two sites. It's also consistent with that proximal reason we gave earlier for the higher information content at the high-elevation station— it's wetter overall, with a greater frequency of heavy rainfall days, giving a more even distribution of days across the four rainfall bins.

It's exciting that we've been able to establish a link between fundamental principles in meteorology and the seemingly disparate field of communications theory, with a greater diversity of weather phenomena being associated with a higher information content. It gives us a new way of looking at the world. But have we made an exciting new discovery

**4.5.** The snow-capped peaks of New Zealand's South Island are an example of orographic precipitation. Moist air from the Tasman Sea, west of New Zealand and southeast of Australia, is lifted upward by the Southern Alps. This generates precipitation in excess of 4,000 mm annually, supporting temperate rain forests and glaciers. These dramatic landscapes made the area a filming location for the movie adaptations of J.R.R. Tolkien's Lord of the Rings trilogy. Photo: Jacques Descloitres, MODIS Land Rapid Response Team at NASA GSFC.

in hydrometeorology with this result? Maybe, maybe not. We would have to repeat the experiment in many different watersheds to confirm the generality of the result. It seems unlikely that there would be some easy, universal, one-to-one correspondence between information content and the number of major storm types affecting an area. And even using the same two precipitation gauges as the basis for the analysis, we would no doubt obtain at least slightly different outcomes depending on our methodological choices. These might include how many bins to divvy up the data into, whether to use hourly or monthly instead of daily time series, whether to consider data only for the wet season, and so forth. Each approach would answer a slightly different question. But we can also have some confidence in our result and interpretation. The difference in the Shannon entropies between the two sites is, by definition, a consequence of observed differences in the relative frequencies of various precipitation amounts at those two locations. And it's a basic fact of atmospheric science that there are generally differences in the relative frequencies of precipitation amounts between low-elevation and high-elevation locations; and also that these differences typically include greater rainfall at high-elevation locations; and further, that the additional rainfall at high-elevation locations is normally due to orographic lift, an additional physical process that does not occur at sea level. So it's entirely reasonable to interpret our observed increase in the information content of precipitation data at high elevations in terms of the addition of another generating source of meteorological information.

How do these results circle back to hydroecology? Do our findings imply that clouds have more to say to fish at our mountain site than at our coastal site? Indeed, they probably do. But that doesn't necessarily mean that the fish get to hear everything the clouds are saying—or even want to! Some caution is needed here. Our application of information theory has allowed us to achieve one of the tasks we set out for ourselves in the introductory section of this book: viewing watershed science through the lens of physics in order to come up with new ways of imagining questions about rivers, and also as a mechanism for using rivers to understand a little more about physics. But at the same time, care needs to be taken not to take analogies or transplanted ideas too far without making the necessary adjustments. Watersheds modify the hydrological signals that the weather sends, through all sorts of complex and nonlinear processes we'll be talking about throughout much of this book. In other words, watersheds function like a computer, performing complicated calculations on the input (weather) in order to produce an output (streamflow). So in a fuller analysis, we'd also need to think in a more detailed way about the role of the river itself in the communications pathway between weather and fish. But the more broadly significant point is that while ideas like Shannon entropy can be shifted easily and successfully from topic to topic, the interpretation must be context specific. Fidelity and speed of message transmission is the Holy Grail of telecommunications. More, and faster, information is better. In hydroecology, though, we should probably expect something very different: the so-called Goldilocks effect typical of biological systems.

Virtually any species harbors a strong preference for an optimal set of environmental conditions, just like Goldilocks wanted her porridge not too hot and not too cold, but just right. Here, that might imply an intermediate, ecologically optimal amount of hydrometeorological information transmission: not too much, not too little. Why? Too little hydrologic variability—and the extreme would be no rainfall or streamflow variability, that is, the zero-information case we discussed when we were introducing the idea of

Shannon entropy above—would represent exceedingly poor aquatic habitat. Fish, like people, need a little variety. Recall those examples we provided at the start of this chapter: fish reproduction is often tied to seasonal flow variability, for instance. As an analogue, we can sometimes see the ecologically detrimental effects of inadequate hydrometeorological variability, that is, insufficient information content, below certain dams that are operated so that they maintain a steady downstream outflow. On the other hand, too much hydrological variability—and the extreme case would be the maximum-information scenario we also discussed when introducing the idea of Shannon entropy above—is deeply problematic for fish as well. Again like people, fish don't like excessive amounts of change and uncertainty; they need a little stability. Excessive sudden flooding can excavate and destroy salmon eggs, for instance, or cause erosion that makes the water too muddy for some fish species. As an analogue, we can see the ecologically detrimental effects of too much variability, that is, of excessive information content, in urbanized watersheds with their increased day-to-day flow variability, a subject we discussed in our chapter on how rivers remember. So we can hypothesize, then, that from a hydroecology perspective, there should be some optimal Shannon entropy in rainfall. The value of that optimum is likely to vary between species depending on their physiological requirements; a Venezuelan piranha and an Alaskan salmon are very different creatures with very different needs—including, perhaps, very different information needs. Clouds should talk to fish, but probably not too much, and each fish is likely to have its own preference.

Where will all this lead? The information age is in many respects still in its infancy. No one knew that the clockwork view of the universe would eventually provide the inspiration and intellectual framework for Newton's physics. We can only wait and see how the communications and computing view of the universe will ultimately impact fields like meteorology, hydrology, and biology. Perhaps someday, we'll have a comprehensive, communications-based theory of watershed biohydrometeorology—of how clouds talk to fish. But we need to start somewhere, and perhaps that comprehensive theory will even be developed by one of the readers of this book!

# 5

### SEARCHING FOR BURIED TREASURE

*Pirates of the Caribbean. Treasure Island. National Treasure. Raiders of the Lost Ark.* Even *The Maltese Falcon* and *The Celestine Prophecy*. All about lost treasure, often buried. For whatever reason, few things capture our imagination and make better stories than the search for buried treasure.

These stories are entertaining, but how plausible are they? Is it all pure fiction? Not entirely. In fact, searching for buried treasure is what exploration geophysicists and geologists do for a living. And by briefly considering what these folks do, we open a door to a wider conversation: about the relationships between mineral wealth and environmental wealth, including the geophysical tools used to investigate both; an examination of perhaps the greatest buried treasure of all—the groundwater that sustains both river flows and the water needs of much of the global population; and an introductory tour of calculus, the universal language of science and engineering.

But first, let's return to our treasure-hunting geoscientists. Some of that work is actually as exciting as it sounds. Getting flown to work in the morning by a bush pilot in a helicopter skimming the treetops, dabbling in international high finance, scrambling up remote mountain peaks where few if any have set foot before, pressing massive supercomputers into service, chancing across exotic wildlife. Packing shotguns for the exotic wildlife, because during more serious forays into the wilderness it is prudent to have, if needed, something at one's disposal a little more robust than good sense and bear spray. Unparalleled summer night skies, and sheets of northern lights—and posh downtown office buildings, clad in glass and oak and brass. And even satellites overhead, scanning the globe for hints at what treasures lie beneath the surface. You will not encounter these things in most other lines of work.

Mining and oil exploration has also, at times, been the source of intrigue that is indeed worthy of a Hollywood movie. Consider the Bre-X scandal. The largest mining stock scam ever, it involved the so-called salting of drill core from the Busang gold deposit in Borneo. Salting consists of surreptitiously, and illegally, adding gold to rock samples after they are collected, but before sending them off to an independent lab for assaying. That artificially boosts the apparent value of a mining claim and therefore the stock price of the junior mining company that owns it, in this case an enterprise called Bre-X. The revelation that the largest gold discovery in the history of the world was, in fact, a complete fraud led to the evaporation of billions of dollars of paper value on stock exchanges. Adding mystery to the mix was the apparent death of the project geologist—by jumping out of a helicopter over a remote tropical jungle. And it doesn't end there: several notable discrepancies were apparent in the Indonesian government's official verdict of suicide, and faked gold concentrations weren't the only dubious activities going on at the Busang property, leading to conspiracy theories about murder on the one hand and a faked death on the other. Fictional yarn-spinning would be hard-pressed to match that.

More enduring and profound is the cultural legacy of buried treasure. The social and economic imprint of resource development in early San Francisco, Denver, and Edmonton were so deep that, a century later, everyday things like their sports teams (49ers, Nuggets, and Oilers) are still named after their respective mineral and oil booms. And the cultural significance of mining didn't start with the gold and oil rushes of the North American West. Around 3200 BC, bronze production began in Mesopotamia, requiring the smelting of both tin and copper and their alloying. By definition, the advent of bronze—which had copper's advantage over stone of being stronger, but stone's advantage over copper of being harder, revolutionizing such things as tool and weapon development—heralded the transition from humanity's Stone Age to the Bronze Age. And the significance of mining-enabled innovation continues today as our technologies march forward. Consider the rare earth elements (RREs), which include obscure-sounding elements like scandium, yttrium, and neodymium. These are necessary ingredients for making lasers, cell phones, and medical imaging technologies. RREs also play a central role in alternative energy, like hybrid and electric cars, solar panels, and wind turbines. As the world strives to move away from oil dependence, RRE availability is emerging as a geopolitical issue. About 90% of global RRE production comes from China, and restrictions it placed around RRE exports have been the topic of World Trade Organization deliberations. The situation may dredge up memories of OPEC and reliance on foreign oil. *Plus ça change, plus c'est la même chose.*

It turns out that such things have powerful connections with rivers. In the search for mineral wealth like gold or copper, rivers serve as pathways—transport pathways, exploration pathways, information pathways, mineral deposition pathways, and sometimes, environmental contamination pathways. We'll start by taking a look at each, and then explore how rivers connect to what might, in fact, be the most valuable prize of them all: groundwater, a crucial resource in its own right and an integral part of the water cycle of every river.

In the days before satellites, helicopters, and vast road networks, rivers were highways for exploration and transport.

**5.1.** A mining exploration camp on helicopter approach in the Omineca, a remote region of northern British Columbia, in the early 1990s. The types of geophysical surveying and subsurface imaging performed by our crew here are widely applicable to questions in hydrology as well. Photo: S. W. Fleming.

They provided access to, and thus a focus for exploration in, rugged terrain. Moreover, forests and other vegetation can shield the soil and bedrock beneath, making it difficult to map the geology or to identify promising mineral exposures, but riverbanks can often expose the underlying geology. And not only did rivers provide early mineral exploration routes over much of the world, but where exploration was successful and mines were built, the rivers then became highways for transport of people and goods to and from the mines and the towns that sprang up around them. This process in turn defined settlement patterns and established what continue in many cases to be today's transportation routes. One example of how mines, rivers, and transportation and settlement patterns interconnect is the Rio Tinto in the Andalusia region of southern Spain. Its banks have been mined for a variety of ores over roughly the last 5,000 years by a long series of Mediterranean civilizations that established mining settlements nearby. The ancient Tartessian culture likely used the river to transport mineral products to port, and a Roman trade route—the Via de la Plata, or Silver Way—that connected to the mines of the Rio Tinto now forms part of the E803, an arm of today's pan-European highway system. Rivers and mineral resources are linked to each other, and also connect civilization's present to its deep past.

Many early mineral discoveries were placer gold deposits, and as such bring to mind another role that rivers play in the search for buried treasure: pathways for mineral deposition. What are placer deposits? These mineral concentrations were formed by the movement of water. Somewhere in a watershed, there is a surface exposure of gold-bearing bedrock—perhaps in a hydrothermal quartz vein created millions of years ago by geologic processes related to the plate tectonics that we discussed in an earlier chapter. Whatever the source, weathering eventually results in some of this gold being dislodged from the bedrock. Being highly chemically resistant, gold doesn't just dissolve or react away to nothing, but is instead transported by moving water as dust, flakes, and nuggets to a creek, then down the creek to a larger creek, and so forth. But gold is very dense—that is, it's heavy. So where a creek's waters happen to slow down, the gold drops out of the water to be deposited on the creek bed. Perhaps the high, fast river flows resulting from some subsequent storm will wash the gold downstream again. Or maybe not; more and more gold may accumulate there to form an economically viable placer gold deposit. The deposit may remain underwater, or it may be accessible only during low-water periods, or the stream may change course and the deposit may be left high and dry. The best-known way to mine such deposits is by gold panning, which is again based on the fact that gold is heavy: mix the dirt with some water, wash the lighter components over the edge of the pan, and eventually you're left with the gold. Larger-scale placer mining operations employ other techniques, but they're based on the same principle. Some of these methods were very destructive, such as hydraulic mining in nineteenth-century California. This involved using water cannon to blast potentially gold-bearing streamsides, freeing up placer gold contained in those slopes but also resulting in massive downstream siltation and habitat destruction—we'll return to that in a moment. Some other minerals of interest show up in placer form, such as the infamous "blood diamonds" once extracted from alluvial deposits in Sierra Leone, Liberia, and other nations in western Africa, and which were used to fund civil wars there in the 1990s. Some commercial placer mining continues today, and its environmental or social implications may still carry some controversy, though modern large mining operations have mostly moved from placer and underground techniques to massive open-pit mines, carving the tops off entire mountains to get at the buried treasure within. Much placer mining is now limited to weekend gold panning—still a great way to reconnect with humanity's deep heritage in mining, and with the rivers that form such an important part of that heritage.

Rivers also act as information pathways in the search for buried treasure. That is, a number of exploration techniques attempt to capitalize on chemical information transported downstream by the current. At any given downstream point in a river, the water passing by has "sampled" the whole upstream part of the watershed. In principle, if there is a significant deposit at some upstream point, then those minerals add their chemical signature to surface water running over it or groundwater running through it. And this signal—elevated concentrations of certain dissolved metals, for instance—is carried downstream. One might then use water quality sampling in rivers and their tributaries to explore for various minerals of economic interest, a technique called hydrogeochemical exploration.

And there is yet another type of pathway function performed—unfortunately—by rivers downstream of buried treasure: potential pathways for ecological harm. We'll give two, quite different, examples of how this can happen and what the implications can be. One is the hydraulic mining of streamsides in California, which we mentioned above. This resulted in massive sediment loads in the rivers being mined, affecting aquatic life and downstream agricultural water users. That devastation led in turn to one of the earliest environmental lawsuits: the 1884 decision by Judge Lorenzo Sawyer in the case of *Woodruff v. North Bloomfield Gravel Mining Company*. This lawsuit effectively ended hydraulic mining for gold in California. Political and legal action continue to be front and center in any controversy around mines and water: again, *plus ça change, plus c'est la même chose*.

A second type of environmental concern often associated with mines worldwide is acid rock drainage. ARD is a slightly odd form of pollution insofar as it doesn't involve any kind of chemical spill, but rather a natural process that may be greatly accelerated by mining activities. Many important ores are made of sulfide minerals (see figure 5.2 and explanatory caption), consisting of a metal combined with sulfur. A common example is pyrite, also called fool's gold due to its golden color, which consists of iron and sulfur. When rocks containing sulfide minerals are broken open and exposed to water and air, as they are during the mining process, the metal and the sulfur dissolve. This liberates heavy metals into the environment. It also creates sulfuric acid, lowering the pH of whatever water course the runoff from a mine happens to run into. The effects on fish in particular can be devastating. A big issue around ARD is that because it is the acceleration of a natural process rather than a discrete chemical spill, it can't just be mopped up. That is, as long as sulfide mineral surfaces remain exposed to air and water, the generation of acidic, heavy-metal-laden water may continue—for generations or even longer. One example of ARD is Spain's Rio Tinto, which we mentioned earlier. It has been mined for sulfide minerals on and off since 3000 BC. ARD can occur naturally, and this may be the case for the Rio Tinto—it has been 5,000 years since it was pristine, so no one can really say—but it seems likely that the intervening five millennia of mining have contributed to the problem. In fact, ARD in the Rio Tinto is so severe that NASA astrobiologists, who investigate extreme environments on Earth to evaluate what possible life on other worlds might look like, have studied the area and used it as a test bed for a prototype of a drill system, called MARTE, that may be used in future missions to Mars. The thousands of abandoned mines in the Colorado Rockies, which in their heyday played a central role in the settlement and wealth of the state, provide another example of ARD. Many have been producing ARD for generations. A few were assigned Superfund status by the US Environmental Protection Agency (EPA), providing funds for their remediation, but collectively, the old mines of Colorado still present a significant problem. In the summer of 2015, the Gold King Mine produced a massive ARD spill that prompted the governor of Colorado to declare a state of emergency for the affected area. The heavy metal plume moved from Cement Creek into the Animas River, then to the San Juan River, and finally into the Colorado River, a major water supply source for the entire US Southwest, although contaminant concentrations were negligible by that point. The incident produced considerable controversy, as apparently it was accidentally triggered by EPA activities intended to address existing ARD contamination on the property, and also drew attention back to the broader question of what to do about these persistent and still-destructive ARD problems.

So, rivers provide a diverse suite of pathways related to mineral wealth—some good, some bad. But the linkages go further. Much of the knowledge and many of the tools used for discovering mineral riches have also turned out to be valuable for understanding water resources.

Consider exploration geophysics. These are more physics-based ways to search for buried treasure than simply following rivers upstream. Indeed, some interpret the word "geophysics"—literally, the physics of the Earth—far more narrowly, referring specifically to a set of techniques used to image the subsurface. Seismic reflection and ground penetrating radar (GPR) involve generating seismic (essentially, sound) waves or electromagnetic waves, respectively, and then recording how they bounce back up from layers or objects in the ground beneath us. Seismic reflection can be used to create two- or three-dimensional maps of Earth extending to several kilometers in depth and is the long-standing mainstay of the oil and natural gas exploration industry, whereas GPR provides very high-resolution images at much shallower depths, often for environmental or civil engineering investigations. Seismic refraction is based on roughly similar ideas to seismic reflection, except that geophysicists watch for

**5.2.** Hydrothermal vents on Eifuku Volcano, more than a mile underwater, off the shores of Japan. Such "chimneys" are one of the most scientifically fascinating places on Earth. Their hot, dark, chemically corrosive, high-pressure environments seem utterly hostile to life, yet unique biota thrive here. And the mineral-rich fluids that circulate in pores and cracks in the rock can form sulfide deposits. Some of these, after being transported into more accessible locations by plate tectonic forces many millions of years after their deposition, became economically minable ores—but in turn, potential sources of acid rock drainage into rivers. Image courtesy of Submarine Ring of Fire 2004 Exploration, NOAA Vents Program.

how the paths taken by seismic waves are bent by changes in rock density, not unlike the way that a lens bends light rays. It's often used for very deep, crustal-scale studies that look tens of kilometers downward. The seismic waves in such cases may be generated by filling a very large borehole with up to a ton of explosives. Watching one of these sources detonate can be quite exciting, to say the least! These techniques are active-source methods in which one observes Earth's response to a controlled source of energy—like that hole full of explosives.

Passive methods are also important. For instance, Earth's magnetic or gravitational fields can be measured very carefully along survey lines. Tiny spatial variations in these fields are tied to variations in the underlying geology, and combined with mathematical models, they may be used to map out Earth's structure below. Perhaps the ultimate passive-source technique is earthquake seismology, in which geophysicists study the way that seismic waves are generated by earthquakes and how these waves propagate away from the earthquake hypocenter (the actual location of the fault rupture—the epicenter, more commonly referred to in news coverage of earthquakes, is the map location on Earth's surface directly above the hypocenter). While the primary goal there often is to understand earthquakes and their impacts, geophysicists have long used earthquake seismology to learn about the fundamental deep structure of the Earth, including the discovery of a solid inner planetary core surrounded by a molten liquid outer core.

Such established geophysical survey methods, which were largely born out of either fundamental Earth science research or the quest for mineral or oil wealth, have also been increasingly applied to questions of environmental and water resource science. For instance, high-resolution, near-surface implementations of GPR and seismic reflection have been used for practical environmental tasks like tracking contaminants in the subsurface and mapping aquifers. An example is shown in figure 5.3. This GPR survey provides an image of a buried channel of the Puyallup River near Seattle. Long ago, the river ran through this spot. Then materials from a landslide on nearby Mount Rainier filled and buried the channel; the river's main channel shifted to another location. Today, the mud of the floodplain contrasts against the sand from the landslide deposits. This distinction allows GPR to directly image the buried channel. In general, due to their typically coarser sediments (sand rather than mud, for instance), buried channels

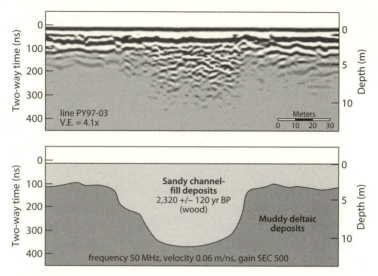

5.3. Ground penetrating radar survey of an ancient, buried channel of the Puyallup River. The illustration runs from bank to bank across the buried river channel; the GPR data are above, and an interpretation is provided below. Image courtesy of the U.S. Geological Survey.

may act as preferential pathways for groundwater flow and contaminant transport, so information like this can be useful in environmental applications, such as remediating polluted industrial sites. And at far larger spatial scales, measurements of Earth's gravitational field by the GRACE satellite pair—fundamentally similar to the gravity survey mentioned above, but performed from space—are used for tallying basin water supply availability, and observing how that might be affected by climatic change.

And these new, environmental applications for geophysical exploration techniques bring us to what is likely the most valuable buried treasure of all: groundwater. About 30% of the world's freshwater lies underground in porous soil and rock. Groundwater may be found in all sorts of places: in the spaces between individual grains of sand or gravel, or in fractures within granite bedrock, or in bubble-like openings within volcanic rock, or occasionally even as underground streams within the caverns of karst limestones. These aquifers individually range in size from relatively compact gravel areas adjacent to rivers, to huge geologic features, such as the Ogallala aquifer underlying several US states. Aquifers contain almost half as much water as all the glaciers and ice sheets combined, and very many times the total amount stored in the planet's rivers and lakes. Apart from the sheer volume and relative seasonal consistency of the resource, an important asset of groundwater as a drinking water supply in particular is that, in general, it is very clean due to the filtration properties of the soil and rock through which it flows. Reliance on this resource is immense. About 40% of America's population and 70% of China's population use aquifers for their supply of drinking water, for example.

Groundwater also plays key roles in sustaining river flows and freshwater habitat. During rainstorms or snowmelt, much of the water infiltrates into the ground to be stored in aquifers. This has two important effects from a river's perspective. The first is that it takes the edge off storm events, reducing flooding and erosion potential. The second is that between storms, and through the dry season, water leaks back out of aquifers into the river, helping sustain surface water flows and habitat. In fact, the distinction between groundwater and surface water is in some sense an arbitrary or even imaginary one. Surface water–groundwater interactions are a fundamental watershed process. Because you generally can't see it, groundwater often doesn't capture your attention the way that rivers do. But aquifers are a fundamental part of rivers: it's sometimes tough to tell where the river ends and the aquifer starts, and you could say a river is simply an aquifer where the groundwater table rises enough to reach the surface. Much of the water flowing in any given river got there only after having passed through subsurface pathways—in most natural watersheds, very little of the water in the river fell as rain or snow directly on the water surface or as other kinds of true surface runoff. Indeed, when the rain falls or the snowpack melts and the river then rises, much of the water in the river isn't the rain that just fell or the snow that just melted: it's soil water that has been around for days, or weeks, or years, and which has now just been displaced by the new, incoming rainfall or snowmelt.

While in some ways more robust than surface water resources, groundwater isn't invincible. Point sources of pollution (such as a chemical spill at a particular location) and nonpoint source pollution (such as fertilizers or pesticides spread over agricultural areas or acres of suburban lawns) can infiltrate downward into an underlying aquifer. Once there, it may contaminate drinking water supplies in the aquifer, or it may slowly but surely flow within the aquifer into hydraulically connected lakes, rivers, wetlands, or the sea. Confined aquifers are groundwater-bearing rock or soil formations that are overlain by one or more layers of impermeable material, which (among other things) offers some protection from such effects. However, unconfined aquifers, which are often shallower and very productive and therefore a common water supply source, lack such protective caps and are typically more vulnerable to contamination. While human pollution is typically the biggest groundwater quality issue, groundwater also picks up the natural chemical signature of the soil and rock it flows through, as noted earlier in this chapter. This could be a good thing in some cases; mineral water, bottled and sold to consumers at a premium, is simply groundwater that contains dissolved minerals from the aquifer in which it resided. In other cases, though, the geology of an area is such that groundwater may naturally experience concentrations of certain chemicals above drinking water standards. Arsenic groundwater contamination, for example, is a huge problem is some parts of the world, such as Bangladesh.

How might one predict the spread of groundwater contamination? This question brings us to the advection-dispersion equation, one of the fundamental equations of mathematical physics. It sees application to a wide range of fields—including other areas of environmental science, such as air quality. Depending on the particular field, it may go by a slightly different name, with "convection" sometimes swapped for "advection" and "diffusion" often switched for "dispersion." This equation employs a class of mathematical techniques collectively referred to as calculus. The knowledge base of science and engineering, from Newton and Leibniz in the seventeenth century to today and beyond, is largely written in the language of calculus, so this is worth exploring.

Calculus essentially involves defining physical processes using a differential equation—an equation containing one or more derivatives, which we'll explain shortly—and then setting out to solve that equation. The fundamental notion behind differential equations is that most things in nature vary in a continuous way over space or time. No matter how many horses you've got under the hood, your car doesn't switch instantaneously from 20 miles an hour to 80—it has to move through all the intermediate speeds to get there. Further, you can subdivide these speeds into smaller and smaller increments—the difference between 70.5 and 70.6 mph can be split into the intervals 70.5 to 70.55 mph and 70.55 to 70.60 mph—over and over again, an infinite number of times. That's what we mean by "continuous." And in practice, what it all comes down to is that differential equations work by defining continuous rates of change.

Let's examine some of these fundamental ideas in science and mathematics by taking a brief look at a relatively simple form of the advection-dispersion equation as it might be applied for estimating the fate and transport of aquifer contamination at a chemical spill site:

$$D\frac{\partial C^2}{\partial x^2} - v\frac{\partial C}{\partial x} = \frac{\partial C}{\partial t}$$

The letter $C$ stands for concentration of the contaminant in groundwater. The $x$ stands for distance (away from the source of the spill, typically), the $t$ stands for time (since the start of the spill, typically), and the $v$ stands for the velocity at which the groundwater is moving. And what about that squiggly symbol, $\partial$? That's essentially shorthand for "derivative," and $\partial C/\partial x$ is called "the derivative of $C$ with respect to $x$." A derivative is that continuous rate of change we mentioned above. So what $\partial C/\partial x$ means is the rate of change of concentration with respect to distance, holding all else constant; that is, how sharply concentration varies from one location to the next at a given time. The higher the value, the more abruptly concentration varies from point to point. Similarly, the derivative of concentration with respect to time, $\partial C/\partial t$, signifies how quickly concentration is changing over time, at a given location. And that $\partial^2 C/\partial x^2$ term? That's a second-order derivative, the derivative of a derivative: it signifies the rate that the rate of change with respect to distance changes with respect to distance. Actually, it's not as convoluted as it might at first seem. In fact, we're all familiar with the concept whether we realize it or not. Speed is the rate at which location changes with time—the derivative of location with respect to time. So what's acceleration, then? It's the second derivative of location with respect to time: how quickly our speed changes. Whenever a sports car ad lists a 0–60 statistic, what they're really doing is giving you a value for a second derivative! $\partial^2 C/\partial x^2$ is a similar idea, except we're thinking instead about concentration and distance downstream from the source. It measures how much the plume of contaminant spreads out as it moves along, kind of like how a drop of milk into your morning coffee spreads out if you just let it sit and watch it. The letter $D$ is a dispersion or diffusion coefficient, and together, $D\, \partial^2 C/\partial x^2$ describes how quickly and how far this "spreading out" happens.

There seems to be a certain irony to such equations. On the one hand, they are an extremely powerful and compact way of looking at nature. On the other hand, in practice much of the elegance associated with them often goes straight out the window, because closed-form analytical solutions to them—which in this case would entail a clear mathematical expression, derived by pencil and paper, giving the concentration, $C$, at any time or location—exist somewhat infrequently for real-world applications. One common reason behind this is that the parameters, like $D$, often vary haphazardly throughout the aquifer. Earlier, we saw a GPR image illustrating one type of heterogeneity in geologic materials across an aquifer. Such variations—say, in the grain size of sediments, or how interconnected the cracks and pores are within bedrock—deeply influences the fluid dynamics of water moving through that geological medium. Unfortunately, although we can map these spatial variations in aquifer properties using technologies like GPR, they can be difficult to describe mathematically. And if we can't assume constant values or a simple mathematical expression for quantities like $D$, then the advection-dispersion equation becomes a mathematical problem that even the world's brightest mathematician couldn't sit down and solve with pen and paper by sheer skill.

By no means does this imply that the equation is no longer valid or useful. Rather, it leads to a clumsier approach to generating solutions: brute-force techniques involving highly repetitive operations in software. However, there is, perhaps, a different kind of elegance to these so-called numerical solution methods, having names like "finite-difference" and "finite-element" techniques. In some sense, these computer algorithms take to heart an aphorism I've heard attributed to Henry Ford: nothing is particularly hard if you divide it into sufficiently small jobs. These methods essentially take the differential equation and replace it with a huge number of very basic arithmetic calculations performed on a grid or mesh pattern that covers the time and space scales of the problem immediately at hand—in our case, the aquifer being studied. In so doing, virtually any applied problem can be solved. Such numerical methods for solving differential equations are prolific throughout the sciences and engineering. Examples range from numerical weather prediction, to modeling the physics of the sun, to designing bridges and cars.

So that's a bit about groundwater quality—how about groundwater quantity? Excessive extraction is another type of problem that can affect the sustainability of groundwater supplies. If groundwater pumping for drinking water or other purposes exceeds the recharge rate from infiltration of rain or snowmelt for example, the amount of water in the aquifer will decline, and over time water levels could diminish to the point that the aquifer is no longer usable as a water supply source. Also, excessive pumping at one groundwater well can lower the groundwater elevation over a wide area, affecting the performance of other wells. This is called an interference effect, and it has led to more than one argument between neighboring homeowners! And even modest extraction rates at wells located near marine environments can result in saltwater intrusion. Another issue that can arise with heavy extraction of groundwater is land subsidence. Consider the San Joaquin River. Stretching nearly from Los Angeles to San Francisco, it forms the southern half of California's Central Valley, which produces about 25% of the food consumed in the United States. A variety of water sources are used to support this vast agricultural production. Groundwater is an important part of that mix, especially in drought years when surface water resources fail. However, extensive long-term pumping of huge volumes of groundwater has dried out the soil, which then settled out and compacted—to such an extent that the surface of the land actually dropped, by dozens of feet in some areas.

Perhaps the most drastic sustainability issue around groundwater, however, lies with so-called fossil groundwater. Aquifers of this type lie in regions that are now relatively arid, and the groundwater within them can be hundreds or even thousands of years old, the leftovers from recharge that occurred during a prior, wetter period of Earth's climate history, such as the last ice age. Strictly speaking, virtually any extraction from such an aquifer may be unsustainable, and certainly, careful water management is key to ensure that the resource lasts. A classic example of fossil groundwater is the aforementioned Ogallala aquifer. The large-scale tapping of this aquifer played a role in ending the infamous Dust Bowl of the American Midwest in the 1930s and ultimately transformed a nearly half-million-square-kilometer area of the semiarid Great Plains (covering parts of South Dakota, Wyoming, Colorado, Kansas, Oklahoma, New Mexico, and Texas, as well as almost all of Nebraska) into one of the most agriculturally productive regions in the world. Unfortunately, this water was originally sourced from recharge during wetter times many thousands of years ago and is simply not being replaced—it's like withdrawing money from a bank account but not making any deposits. The resulting groundwater "mining" has resulted in many wells going dry; the Ogallala as a whole could dry up one day not too far in the future, though steps are being taken now by some to adapt.

Earlier, we referred to groundwater as buried treasure. That wasn't just a metaphor. The sums of money involved can be astronomical. For starters, there are the losses associated with improper management and protection of groundwater, what economists call externalities: lost recreational or commercial fisheries opportunities linked to depleted but ecologically necessary summertime groundwater inputs to streams, say, or the health care costs associated with treating people affected by polluted well water. And that's not even to mention follow-on legal and political issues, and all the associated costs. But financially speaking, it goes far beyond questions of environmental sustainability or public health, as important as those obviously are, to basic relationships between water, land, and money. My first foray into environmental and water resource consulting was in Colorado in the late 1990s, and by that point there was already a well-established water market of sorts in the state: farmland was being snapped up for the associated aquifer rights, with the water then sold commercially. The water below ground was worth more than the real estate above. Water has of course been a big deal in the mostly dry American West for a very long time, and it's an even bigger deal now—and increasingly everywhere else as well. To put some numbers on it, the National Ground Water Association reckons that the total economic value of pumped groundwater is over $20 billion (yes, that's billion with a "b") every single year, and that's just in the United States alone. Buried treasure indeed!

Remember how we started this chapter with the names of popular movies themed around the search for bounty? Now we can add groundwater to the list of Hollywood-sanctioned buried treasures: the 2008 Bond flick, *Quantum of Solace*, wound up being about exactly that. Well, that—and Aston Martins, Tom Ford suits, gorgeous Bolivian secret agents, and impressively high-end bartending. Good company to keep.

# 6

## THE DIGITAL RAINBOW

Consider a prism, like that on a good-luck charm. A beam of white light goes in. Out the other side comes a whole spectrum of color, the complete rainbow. Light is electromagnetic radiation, which can be viewed as waves of energy, and it covers a range of wavelengths or, more or less equivalently, frequencies. Frequency is simply how quickly something varies. Within visible light's frequency band, red light has the lowest frequency—its energy waves bob up and down slowly, like the mellow wash of waves on a quiet beach—and violet the highest, perhaps more like turbulent splashes in a fast-flowing river. When lights with all these different frequencies are mixed together in equal proportion, they form white light. A prism just reverses the blending process, splitting the white light into its constituent colors or frequencies.

Essentially the same concepts apply to anything that varies in time—in other words, a time series, such as daily measurements of river flow. Any time series can be viewed as being kind of like white light, that is, a blend of individual signals having different frequencies. And again, the blending process can be reversed by a prism, or something very much like a prism. That decomposition of a time series into its frequency spectrum is called spectral analysis. There are different ways to do it. The classical (and still highly effective and probably most common) one is based on the ideas of the eighteenth- to nineteenth-century mathematician Joseph Fourier, and in particular a mathematical operation called the Fourier transform. Spectral analysis of real data sets involves numerical algorithms implementing mathematical techniques like the Fourier transform but specifically adapted to or designed for the computer, hence the statistician J. D. Brillinger's suggested phrase for the numerical spectrum: the digital rainbow.

Fourier decomposition and spectral analysis are fundamental techniques of mathematical and statistical physics, and as we'll see, these "digital rainbows" can reveal much about river flow variations, from the damping of day-to-day fluctuations, to the regular march of the seasons, to longer-term events and shifts related to climate. Let's start with a simple example of how spectral analysis can be applied to river flows and what it can tell us. We briefly considered the Congo River in chapter 3. The second-largest river in the world after the Amazon, it sprawls across nearly a dozen African nations, and its massive upstream basin area includes parts of both the northern and southern hemispheres, such that it can be winter in one part of this huge watershed while it's summer in the other. Complexities like these result in an average of two seasonal peaks per year at downstream locations like Kinshasa: a major peak around December, give or take a month, and a second, distinct, but much smaller one around May. Twin seasonal peaks can also occur in rivers elsewhere for different reasons. In parts of the Pacific Northwest, for instance, precipitation experiences a single wintertime peak, but it gets split between rainfall at lower, warmer elevations and snowfall at higher, colder elevations. As a consequence, rivers in this area can experience rain-driven high flows in winter, and then snowmelt-driven high flows in summer, with periods of lower flows in between these two seasonal peaks. Very similar effects occur in parts of Scandinavia. Alternatively, some areas experience two seasonal precipitation peaks—in northern Florida, for example, there's a strong summertime rainfall maximum from thunderstorms, and another, minor peak in late winter or so associated with frontal storms. This again can give two corresponding seasonal peaks in streamflow.

The more common situation, however, is just one yearly peak in streamflow. As an example, consider the Thames. Although not a particularly large river by global standards, the Thames holds a distinctive place in history and culture as the heart of London, and many British landmarks lie within sight of it. While Celtic tribes were present beforehand, it's believed that the city of London was originally founded by the Romans about two thousand years ago. Site selection was guided by the Thames; it facilitated access to Europe, and they chose a point on the river that was narrow enough to bridge. The Thames is also a hydraulically interesting river. Ocean-emptying rivers like the Thames experience backwater effects due to marine processes like tides and storm surges, which raise and lower the water level at the mouth. What this means is that, occasionally, the water level at the mouth can increase so much that the usual slope of the water surface reverses, and at such times, much of the lower Thames flows "uphill" from the sea toward the headwaters. That said, the hydrologic seasonality of the Thames is relatively simple; it has (on average) a single annual peak in wintertime, generally sometime between December and March depending on the particular year, reflecting a single annual peak in rainfall. This is common for many rivers worldwide, such as the summer monsoon-influenced timing of rivers like the Brahmaputra in India and Bangladesh, or the spring snowmelt-controlled timing of rivers like the Colorado in the United States and Mexico.

To recap: on average, the seasonal flow cycle for the Thames has one peak, whereas that for the Congo has two peaks. And that happens every year, the once-per-year seasonal cycle in the

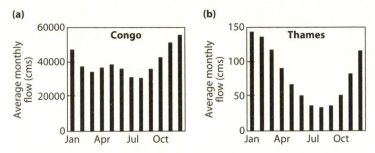

**6.1.** Average annual flow regimes of the Congo (a) and Thames (b) rivers.

Thames and the twice-per-year seasonal cycle in the Congo, repeating itself year after year after year, almost like clockwork. In other words, Thames flows oscillate with a dominant frequency of one cycle per year, and Congo flows vary at a frequency of two cycles per year, though one of those seasonal peaks is bigger than the other.

So how does spectral analysis work, and what can the resulting digital rainbow tell us about these two rivers? Although there are different ways of doing it, the basic premise is more or less along the lines of fitting a whole bunch of sine waves to the time series. A sine wave is just a type of regularly and smoothing oscillating signal. The sine operator is the same one you may or may not remember from high school geometry—it has to do with the relationships between the lengths of the sides of a right triangle, but you don't need to worry about that here. Examples of a few different sine waves are shown in figure 6.2. Key points to note are that sine waves can have different amplitudes, frequencies, and phases. The effects of these three parameters are also illustrated in the figure. The amplitude sets the size of the oscillations, the frequency is how quickly the wave oscillates, and the phase takes the whole signal and shifts it forward or backward in time. By combining individually simple sine waves having different characteristics, we can build a complicated and

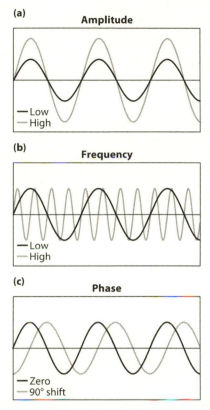

6.2. In each panel of the figure, the sine wave in black is identical; the gray sine wave illustrates the effect of changing the amplitude (a), frequency (b), or phase (c).

irregular-looking time series. Spectral analysis does the opposite: given a complicated and irregular-looking data set, spectral analysis decomposes it into its simpler constituent parts, like those shown here, identifying the sine wave frequencies present. This can be very useful for discovering patterns and causes in complex geophysical data sets, such as streamflow observations.

That is, the basic idea behind spectral analysis is to take a number of different sine waves, each having its own amplitude, frequency, and phase, and then add them together to recreate your original time series (such as river flow data). The trick, of course, is to determine how many sine waves you need and what their individual characteristics need to be. Once that's done, though, you can look at how important each of the sine waves is, as a function of frequency. We do this by making a graph—the numerical power spectrum—that essentially shows how large the amplitude is for each value of frequency. If there are no important regularly timed oscillations in your data set, you get a so-called white noise spectrum, which is a power spectrum that's basically flat. The name comes from the fact that all frequencies have equal weight, like in white light. Remember that prism we mentioned at the start of the chapter? Each color is present in more or less equal amounts in white light, so when a prism breaks up white light into its constituent frequency components, all the colors of the rainbow are there. On the other hand, if there are one or more important oscillations in the time series, we instead get peaks in the spectrum corresponding to the frequencies of those oscillations. Another way of saying this is that there is a lot of signal power in those frequency bands. Continuing the analogy to light, in this case not all the colors of the rainbow would be present, or present in equal amounts—one or more colors would stand out above the rest.

Let's return now to how this works in hydrology. For the Thames, where we have a big seasonal oscillation producing one peak per year corresponding to the annual winter freshet, we simply get one big spike in the observed power spectrum at a frequency of one cycle per year (see figure 6.3). For the Congo, we instead have two distinct seasonal streamflow peaks. Remember from our earlier graph of the average flow regime that there's a dominant annual cycle with a maximum that repeats every year around December, but also a second smaller peak around May, which appears as sort of a "blip" superposed

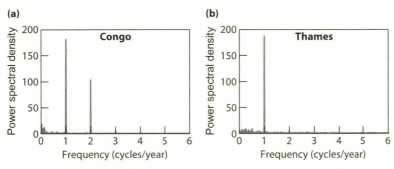

6.3. Power spectra—digital rainbows, as it were—of monthly streamflow time series for the Congo and Thames.

on top of the dominant annual cycle. So the observed power spectrum shown here for the Congo River reveals two spikes: a big one at a frequency of one cycle per year (identifying the major December-peaking annual cycle), and a smaller one at a frequency of two cycles per year (telling us that there are, in fact, a total of two seasonal peaks every year—including that second, minor peak in May).

That's all fine and well, but the spectral analyses haven't yet revealed anything about flow variations in the Congo and Thames that we didn't already know. Can we get any more information out of this technique? The answer is of course yes. But to do this, we'll have to delve into something we'll call spectral signatures.

Up to now, what we've been looking at here are variations that happen on pretty much strictly regular time intervals, like the march of the seasons over the course of a year, repeating itself year after year. Signals like that are called periodic. A relatively small number of sine waves—in the simplest case, just one—can be used to represent such signals. And the resulting power spectrum consists of a handful of sharp spikes, as we saw above for the Congo or Thames. It turns out, though, that one of the remarkable features of the Fourier transform is that you can use a combination of different sine waves to represent any time series, even if it isn't periodic. The trick is to use enough sine waves—in principle, an infinite number of them. The end result is that two different ways emerge for representing a time series. One of these is called the time domain: simply the familiar graph showing measurements of streamflow on each successive day, for instance. The other is called the frequency domain, and consists of both a power spectrum and a phase spectrum. The power spectrum is what we looked at above, and we'll be ignoring the phase spectrum here. The take-home point is that the power spectrum captures all sorts of information that doesn't have anything to do with clockwork oscillations like the seasonal cycle. And because those types of variations in streamflow aren't periodic, they don't show up as strong, crisp individual peaks in the power spectrum. Instead, they're reflected in the overall shape of the spectrum, often in distinct and recognizable ways that correspond to particular hydrological or climatic processes. These are called spectral signatures.

Looking at the power spectra for Congo and Thames streamflow data shown in the preceding figures, there don't seem to be signatures of anything at all, except for the seasonal cycles. But that's because the march of the seasons is, overall, by far the largest type of variation that most rivers will ever experience. The seasonal cycle has huge signal power—so much so that it overwhelms everything else. As a result, you often have to tackle the spectral analysis problem in a slightly different way to see subtler, though still very important, phenomena.

One way to accomplish that is to use double-logarithmic plots. For our purposes, just consider this an alternative way of drawing graphs. On a logarithmic graph, the distances between consecutive tick marks on the graph axes are in multiples of 10, so that the distances between 0.1 and 1, between 1 and 10, and between 10 and 100, are all the same. Logarithmic plots tend to damp out the visual importance of large data values and to boost the visual importance of small ones. The net result is that big, obvious things (like the annual or semiannual spikes in the foregoing power spectra) are suppressed a bit. Instead, subtler features receive more emphasis, like the spectral signatures we want to look at.

Both a time series and a double-logarithmic plot of the power spectrum are given below for the Columbia River. The Columbia is, by flow volume, the largest river on the western coast of the Americas—the greatest Pacific drainage basin in the Western Hemisphere. It covers parts of seven US states and one Canadian province, with transboundary river management coordinated under international treaty. Its basin area spans icefields on the western slope of the Rocky Mountains, the temperate rain forests of the Pacific coast, the soaring volcanoes of the Cascades, a high desert in the rain shadow of the Cascades that continues uninterrupted to Mexico, and productive agricultural valleys like the Willamette and Okanagan—not to mention some of the greatest salmon habitat to be found anywhere, at least before the dams came. The Columbia is also a heavily managed river, with dozens of major dams variously operated for hydroelectric power generation, flood control, irrigation water supply, and so forth. An unfortunate side effect is the disruption of the region's once-stunning salmon runs: hydropower is clean energy, but it's not necessarily green energy. The power generated by dams on the Columbia is monumental, with an annual value in the neighborhood of a billion dollars, keeping the lights on from California to British Columbia. It also played a surprising role in changing the nature of warfare forever. The Hanford site used abundant hydroelectric energy afforded by the Columbia's Grand Coulee and Bonneville Dams to produce plutonium for both the Trinity atomic bomb test and the Fat Man device detonated over Nagasaki, ending World War II and ushering in the nuclear arms race of the Cold War. Although there is considerable variation among its tributaries,

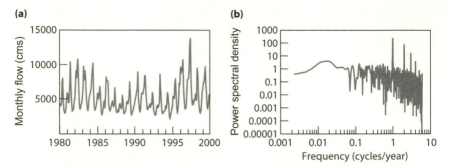

6.4. Snapshot of monthly flow time series for the Columbia River (a), and double-logarithmic plot of the power spectrum obtained from all available data (b). Although the largest peak in the spectrum still corresponds to the once-per-year freshet, this alternative data visualization also reveals the spectral signatures of subtler dynamical features.

with rainfall and glacial ice melt being important along the coast and in the headwaters respectively, the overall seasonal hydrology of the Columbia is dominated by spring-summer melt of the seasonal snow pack, giving a relatively straightforward once-per-year cycle.

One visual signature of the Columbia spectrum is immediately clear. While we still see a few spikes, such as that corresponding to the once-per-year cycle, the spectral response is generally flattish but high on the left-hand half or so of the power spectrum graph (that is, at lower frequencies), and then starts sloping downward on the right-hand side of the graph (at higher frequencies). This overall pattern is often called a "red noise" signature, by analogy to the visual light spectrum: red light consists of low frequencies, somewhat like how the log-space power spectrum of the Columbia River is dominated by low frequencies. We can also make an analogy to sound waves. Lower frequencies are like a bass or cello, whereas higher frequencies are like an electric guitar or violin, so that red noise is a bass-dominated noise that could be imagined as a deep rumble, while white noise is more like a hiss.

For river flows, this effect arises largely from the watershed's ability to store water. Such storage mechanisms can take the form of porous soil and rock, wetlands, lakes, and so forth. But a very clear example that you can directly experience yourself is a process called canopy interception storage. Stand under a big leafy tree as a rainstorm begins. You'll be kept dry. The falling rain gets caught and stored up in the tree. But if the storm continues, within a few hours that canopy of overhanging branches and needles won't keep you dry anymore; the rainwater starts dripping through, right onto you, because the canopy's ability to store water has been exceeded. And after the storm stops, rainwater will often continue to slowly drip down from the canopy and run down the trunk of the tree, depleting the interception storage. Now multiply this effect by all the trees in the forest, and you can begin to see what's going to happen. For the rain distributed across the watershed to get to a creek channel, it first has to make it through the forest canopy (and then overland or underground to the riverbank). But various terrestrial hydrologic storage mechanisms, like canopy and aquifer storage, grab the water, hold on to it for a bit, and then gradually release it. So the immediate effect of highly variable, fast-changing (in other words, high-frequency) weather processes—in this example, rain showers that come and go—is damped. But on the other hand, much of the rainwater from that storm is still going to eventually make it to the river, just in a slower, steadier (in other words, low-frequency) fashion. Hence, red noise: attenuated power at high frequencies, enhanced power at low frequencies. Another way of looking at the effects of watershed storage is to view it as memory: river flows reflect not only the weather right now, but also the weather over the preceding hours or days, because the effects of past storms are still trickling out of storage into the river channel. Thus, more generally, red noise is the spectral imprint of system memory: the system remembers what's come before, a subject we touched on in chapter 3.

Another thing we see in the logarithmic plots is a series of moderate, somewhat indistinct peaks at frequencies lower than one cycle per year, that is, corresponding to effects that occur only once every few years. Unlike cycles related to the seasons, which show up in the power spectra as large, crisp peaks, these slower variations appear as broad or vaguely defined highs. You have to be a little careful when looking at some of these features: researchers have "discovered" a lot of cycles in everything from stock prices to species extinctions on the basis of overenthusiastic interpretations of power spectra, only to find out later that they were just chasing ghosts. The basic problem is that the variation of interest may be subtle, and the data set may not be very long compared to the timescale of the phenomenon—spectral analysis works best when the data record is long enough to capture many repetitions of the cycle. But with the benefit of independent lines of evidence using other techniques and data sets that we won't go into here, we can say with confidence that some of the lower-frequency peaks in the Columbia power spectrum are the imprint of large-scale ocean-atmosphere circulation patterns. These patterns are also known as climate modes, because they are coherent, organized modes of climatic variability. One circulation pattern that significantly influences Columbia River streamflow is the Pacific Decadal Oscillation (PDO), which shows up in the spectrum above as a wide peak centered roughly around 0.02 cycles per year. A low-frequency phenomenon of the northern Pacific Ocean, it involves a switch between "cool" and "warm" phases every few decades, deeply affecting climate across much of northwestern North America. The PDO has impacts on issues ranging from water resources to forest fires to skiing conditions to fisheries resources. Indeed, abrupt, long-term shifts in salmon returns were key to identifying this climate pattern. By far the best-known climate mode, though, is El Niño–Southern Oscillation, often abbreviated ENSO, which has its origins in the tropical Pacific Ocean.

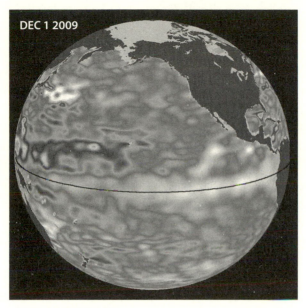

DEC 1 2009

6.5. Satellite image of the 2009–10 El Niño event, appearing as a band of warmer-than-normal sea surface temperature along the eastern equatorial Pacific, reaching to the South American coast. Photo: Jet Propulsion Laboratory.

ENSO more or less consists of three states: El Niño (also called the warm phase), La Niña (cold phase), and a neutral state. The term "El Niño" (Spanish for "The Child," referring to the Christ child) was coined for this phenomenon by fishermen off the west coast of South America centuries ago, as these events affected the fishing and tended to occur around Christmastime. El Niño conditions involve a weakening of the easterly (that is, from the east and toward the west) trade winds. These winds normally pile up warm surface waters in the western part of the Pacific Ocean, close to Asia and Australia. When the winds slacken (or even reverse) during an El Niño, those warm surface waters flood eastward across the Pacific—in fact, all the way back to the west coast of

South America. La Niña is roughly the opposite state of affairs. The entire process is extremely complicated and still not that well understood. ENSO involves nonlinear, possibly chaotic dynamics; random forcing; and complex interactions and feedbacks between the ocean and atmosphere on a hemispheric scale. Still, the foregoing is a loosely correct description of what an ENSO event looks like.

How can a pool of warm water concentrated in the western Pacific flood all the way back, across the entire ocean, to coastal South America, without completely dissipating first? This is made possible by a remarkable phenomenon called the equatorial waveguide, which in turn results from the Coriolis effect. The Coriolis effect has to do with the momentum of something passing over the surface of the rotating Earth, plus the fact that the eastward absolute velocity of the Earth's surface decreases from the equator northward or southward, reaching zero at the poles. At any latitude, you're going to experience one rotation of the Earth per day. But the actual distance you have to cover in that day is a lot bigger at the equator simply because the Earth is bigger there—or more specifically, the distance between the Earth's surface and its axis of rotation is largest there. In contrast, at the poles you're just kind of spinning in place like a figure skater doing a pirouette. The ultimate consequence of the Coriolis effect is that everything from ocean currents to missile trajectories get pushed to the right (relative to Earth's surface) in the northern hemisphere, and to the left in the southern hemisphere. The Coriolis effect also plays a role in the spiral-arm dynamics of hurricanes (in the Atlantic and Caribbean; called typhoons in the western Pacific, and cyclones in the Indian Ocean) that you see on satellite images on the TV news, for instance. The important bit for the so-called equatorial waveguide is that when a wave of warm water moves eastward across the equatorial Pacific, if it begins to stray northward into the northern hemisphere it gets pushed to the right (back south), and if it starts to stray southward into the southern hemisphere it gets pushed to the left (back north). The wave is thus guided along the equator. The process is analogous to the waveguide that operates within fiber optic cables, for example; light straying from its down-cable path is bent back toward the center. That's why fiber optic cables are so efficient at transmitting data.

And how does ENSO, essentially a tropical phenomenon, makes itself felt in the Pacific Northwest's Columbia Basin, and indeed worldwide? El Niños and La Niñas ultimately wind up affecting climate on a planetary scale through a phenomenon called teleconnectivity—literally, "remote connections." This happens mainly by altering huge waves in the atmosphere, known as Rossby waves, which appear as meanders in the jet stream you might see on the nightly TV weather report. The resulting teleconnection patterns can be somewhat intricate. For instance, ENSO impacts on precipitation are opposite between the Pacific Northwest and the Alaskan panhandle, and again between the Pacific Northwest and the American Southwest, with intervening "hinge points" of little or no effect somewhere in the vicinity of the central British Columbia coast and northern California respectively, whereas the impacts of ENSO on temperature are more or less uniform across that whole region. One teleconnection that is relevant to the flow of the Columbia River is that, by altering the paths of incoming storm tracks off the Pacific, an El Niño event will tend to give warmer- and drier-than-normal winters across much of the basin. Such climatic conditions tend to yield lower snowpack in the Pacific Northwest, as we became keenly aware of during the 2010 Winter Olympics, forcing extensive use of snow-making machines. Roughly the opposite happens during La Niña years.

That's the broad story about ENSO, but there's a lot of fine print. For instance, the effects described above are only "on average." ENSO is just one source of variability in weather and water. Sometimes during a La Niña winter, for instance, you may end up with weather that's different from what you'd normally expect from a La Niña. Such less-than-complete consistency has practical implications. For instance, it limits the degree to which seasonal river forecasting systems, such as those operated to support hydroelectric power generation and water supply planning, can benefit from including information about ENSO state. Another way of saying this is that, when it comes to climate oscillations like ENSO, the signal may not always be clear and strong. There are still more subtleties: recent work suggests that ENSO impacts on precipitation and temperature can be highly nonlinear—specifically, parabolic—in some regions. We discussed nonlinearity in chapter three. What this means for water supply teleconnections is that in these particular regions, instead of being opposites, very strong El Niños and very strong La Niñas may have impacts that are similar to each other—but different from neutral conditions! Implications for river flows are still being assessed, but the results so far suggest that it may actually provide ENSO-based seasonal water supply forecast skill in areas where little or none had previously been found, such as the Sacramento River of northern California. When you start digging into the details, things can get a bit complicated, but also interesting and potentially valuable.

The power spectrum corresponding to ENSO is a little complicated, because ENSO events are quasi-periodic. This means that, on the one hand, they don't happen at the regular, constant intervals that periodic processes (like the seasonal cycle) do. On the other hand, their timing isn't totally haphazard, either; it's not like you'll go fifty years without one. El Niño–Southern Oscillation has timescales of about two to nine years or so—in fact, it has all these timescales, though not in equal proportion. A spectral analysis of an index of ENSO conditions will reveal a kind of spectral "signature"—a series of peaks spread across frequencies of roughly one cycle per two or three years to about one cycle per eight or nine years. And these sorts of spectral signatures show up, in turn, in our spectral analysis of the Columbia River.

We've talked about this series of spectral peaks associated with year-to-year and decade-to-decade climate oscillations, as well as the red noise imprint that watershed storage and memory leave behind on the overall form of the power spectrum. But it turns out there are other kinds of climate variability and change in addition to oscillations like ENSO and PDO, and that some of these can produce red-noise signatures in the spectrum that look a lot like those associated with memory. Consider a gradual upward or downward tendency in a time series, such as a progressive rise in temperature or general decline in streamflow—in other words, a trend. Just what exactly constitutes a trend can sometimes be a matter of perspective. Say we have a cycle that takes a very long time to complete. If the observational time frame is short compared to the length of that cycle, then the rising limb of the oscillation (for example) would appear as an upward trend, with little hint that it's really just part of a longer-term pattern of undulating values. Conversely, some changes really do involve an underlying structural shift in the system to a new (for instance, warmer, or drier) state. Either way, spectral analysis of a dataset containing a trend produces a red noise signature a lot like that we discussed earlier. This effect is so well known that time series analysts routinely consider it a trivial nuisance in the context of spectral analysis, and process such trends out of their datasets before performing their work, allowing them to focus on the effects of greater interest to them. I've followed this standard practice in the spectral analyses I've shown above. Yet the longevity of the environmental changes associated with these red spectrum-producing trends are such that they are deeply important phenomena in their own right.

These longer-term climate changes can arise from a variety of sources, some of which are natural. Indeed, we would be living in a very unusual phase of Earth's history if some kind of climatic change wasn't happening. Variations in Earth's orbit, called Milanković cycles, are one source of climatic change. These played a major role in the waxing and waning of the Pleistocene ice ages. Everything from changes in the sun's output to volcanic eruptions have also been important. The planetary climate system also has complicated internal dynamics, called fractal processes, which produce long-term trends in one direction (say, toward warmer temperatures) but then abruptly reverse course. We'll broach many of these natural effects in later chapters. But human-induced, or anthropogenic, climate changes are also increasingly important. One such impact is the urban heat island; replacement of forest and grassland by large concentrations of concrete, asphalt, and heated buildings in major cities typically raises the temperature locally, or even regionally.

Unfortunately, another anthropogenic impact appears to be truly global in scope and accelerating—and that, of course, is the effect of greenhouse gases. Greenhouse gases include carbon dioxide, methane, and water vapor among others, and they act to trap heat at Earth's surface, increasing the temperature. Short-wavelength solar radiation passes through the atmosphere relatively unfettered, because these greenhouse gases aren't very effective at capturing it. This sunlight heats the surface, which then attempts to cool off by emitting long-wave radiation. But atmospheric greenhouse gases are pretty effective at absorbing such long-wave radiation. They then re-radiate it back down to Earth, raising the temperature. This "greenhouse effect" is a long-established fact of planetary geophysics. It's also perfectly natural and a very good thing. Without a natural greenhouse effect, Earth's surface would be frozen solid, and life as we know it wouldn't exist. But on the other hand, too much of a good thing leads to excessive temperatures. An extreme example is Venus, which has an average surface temperature of about 460°C due to an out-of-control greenhouse effect.

Human society has been artificially increasing the concentration of greenhouse gases in Earth's atmosphere. This has occurred through several mechanisms. Two of the most important are combustion of fossil fuels, which releases carbon dioxide and other pollutants into the atmosphere, and deforestation, which reduces the Earth system's capacity to remove carbon dioxide from the atmosphere. Both are linked to the massive growth of global human populations. In fact, one study demonstrated that personal reproductive choices are the single largest impact that most of us can have on our individual carbon legacies. Understanding the whole picture—including not only fossil-fuel-based energy generation, but also deforestation, population growth, and economic growth—is important, because the balance of evidence, as it is currently understood, points to anthropogenic climate change as the main source of the possibly unparalleled rate of increase in Earth's surface temperature over the last century or so, and suggests that it will accelerate into the future with tremendous consequences. A couple degrees up or down here or there, over the course of decades or generations, may not seem like much. But such a long-lived and widespread change can profoundly affect the dynamics of the atmosphere and oceans, and therefore everything related to them, from the success of commercial fisheries to the amount of energy required for air conditioning. It's important to bear in mind here the extraordinary power of nature. It has been estimated that over its short lifetime, a hurricane can release the same energy as 10,000 nuclear weapons, and just such a single hurricane came shockingly close to wiping New Orleans off the map—a major city in the most powerful, wealthy, and technologically innovative nation that human civilization has yet produced. So the potential for a few more storms per hurricane season, or perhaps fewer but larger storms, as might occur under long-term climatic changes, is not just some bland or academic statistic. And this is just one example of the kinds of practical ramifications that anthropogenic global climate change might conceivably have.

One of the basic tenets of Earth science is that the past is the key to the present, and possibly to the future. It turns out that the geologic record teaches us quite a lot about climatic change. For example, one finding, which ties back to the climatic oscillations we discussed earlier, is that warm periods in Earth's distant past had several characteristics in common with an El Niño event—a more or less permanent one. Indeed, many studies have looked at the way that systems, such as rivers, have responded to El Niños over the historical observational record as an analogue to climatic change. By in some sense bridging the gap between the spectral peaks of short-term climate oscillations and the red spectrum of long-lived trends, these studies can give us some further clues as to what the impacts of climate changes might eventually be.

It's interesting to compare results from such observational studies of ENSO impacts on rivers to outcomes from studies based on general circulation models (GCMs) of global climate change. GCMs began as mathematical and computational models of the coupled circulation of the atmosphere and ocean, but have expanded beyond that to become comprehensive and hugely impressive (but still far from perfect or complete) models of the Earth system. That includes the biosphere—all the living things on the planet to the extent that they may affect the global climate, such as the way forests store carbon. As one might imagine, the computational power involved is immense. These virtual-Earth models are run on massively parallel supercomputers with capabilities that dwarf anything most of us ever come into contact with. That said, given how much may be at stake, it's interesting that an endeavor with the magnitude, gravitas, and ambition of the Apollo Mission or Manhattan Project has not been undertaken yet to conclusively and exhaustively assess the impacts of global climate change and the effects of different mitigation approaches. Work has instead been undertaken semi-independently by an assortment of comparatively small research organizations worldwide. Each of these use their own GCM and focus primarily on publishing results in the academic literature, which is occasionally summarized and synthesized in reports by the Intergovernmental Panel on Climate Change. There are good historical reasons why the field evolved in such a way, and this comparatively noncentralized approach has the advantage that it captures different ways of understanding the problem—an important asset when tackling difficult questions about a complex, open, planetary-scale system like climate. However, the results can vary significantly between these different models, and none of them seems to reproduce past climate behavior with quite the level of detail and accuracy that an applied scientist or engineer would ideally like to see when using model predictions for making major design, management, and policy decisions. Climate projections for the future also vary between the different emissions scenarios used as input to the models. These scenarios are estimates of greenhouse gas emissions, in turn representing an attempt to predict the potential future trajectories of the global economy over the next century—an ambitious venture, to say the least. The climate projections are also different for different regions, reflecting spatial variations in climate. As a consequence, the collective output from all these models are better viewed not so much as forecasts or predictions per se, but instead as scientifically plausible scenarios providing a bracketed range of uncertainty in potential future climate that might in turn be responsibly used to guide our planning.

What does all this signify for the flow of the Columbia River, and how does it compare to the effects of El Niño? Overall, GCM projections call for increased temperatures, and possibly somewhat increased precipitation, across much of the Columbia Basin. The hydrological impacts of those projected climate changes are likely to consist mainly of a shift in seasonality, that is, a de facto transfer of water across the seasons. Specifically, warmer temperatures will probably mean that a higher proportion of the abundant wintertime precipitation in this region will fall as rain instead of snow. This will in turn increase wintertime flow, but also decrease the winter snowpack, such that flows during the springtime melt freshet will suffer a commensurate decrease. Late-summer flows may also decrease, and on average, the springtime melt freshet will also likely take place earlier. By and large, these impacts are qualitatively very similar to what has been observed to happen historically in the Columbia Basin during El Niño years.

And what will be the broader implications of these geophysical changes—not only in the Columbia Basin, but worldwide? The question remains difficult to answer reliably, for a number of reasons. Uncertainty in climate projections, area-to-area variations in climate forcing and hydrologic responses, and lack of clarity on how ecosystems and economies will respond are some of the factors at play. Additionally, it remains unclear how large the aquatic ecological effects of projected climate changes will be relative to other environmental insults, such as urbanization and pollution, or similarly, how significant the water supply implications will be relative to other management issues like aging or leaky infrastructure, inadequate water metering and pricing in some areas, and increasing global water demand under continued population and economic growth. Water resource practitioners need reliable and specific details, so these sorts of uncertainties make it more challenging to prioritize climate changes and plan for them effectively from a water resources perspective. And practical, on-the-ground solutions are not easy to come by. Building more and bigger water supply reservoirs is extremely expensive and commonly has major ecological implications. In fact, the era of major dam building in North America ended, for the most part, decades ago. Stopping climate change is, as we've seen, easier said than done. Moreover, even if tomorrow the whole world instantly ceased all greenhouse gas production, the climate system would

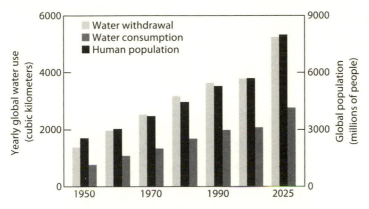

6.6. Global population, and water withdrawals and consumption (lower than withdrawals, because some withdrawn water returns to natural water bodies after use). Data are from the US Census Bureau and UNESCO. Population growth, and the agricultural and industrial expansion it generates, have water supply implications: water pollution, increased water demand, and growth of activities like fossil fuel combustion and deforestation that contribute to global climate changes. Water demand needs to reach only 40% of local available supply to produce extreme scarcity, marked by intense competition for water and major negative impacts on economic development.

continue to respond to past greenhouse gas inputs for decades before coming to a new equilibrium and stabilizing. Curbing population growth in a democratic and socially just manner has proven difficult globally, and perhaps more importantly, many urban centers are aggressively accumulating population from surrounding areas, deeply exacerbating local- to regional-scale water stresses. So-called no-regrets policies—actions that should be desirable regardless of the details of climate change, such as working to improve water use efficiency—seem an obviously desirable path, but even no-regrets policies have a cost, and it seems unlikely they will be enough. Is it finally time for a truly "big-science" approach, for that Apollo Mission or Manhattan Project for climate change that we mentioned earlier? Perhaps it's an option worth considering.

# 7

## LANDSLIDES, FRACTALS, AND ARTIFICIAL LIFE

The storm arrived late in the year—well after the end of coastal Venezuela's usual tropical rainy season. It started innocuously, with a few days of moderately heavy rainfall. But December 14, 1999, marked the start of a deluge that would drop almost a meter of rain in this steep mountain region in only three days— almost as much as some parts of soggy Ireland get in a year. For the inhabitants of Caraballeda, a Caribbean port city through which the Quebrada ("mountain stream") San Julián flows, the disaster came the following night. Eyewitness accounts told of rumbling sounds, crashing rocks, and waves of water. By daybreak most of the damage had been done. Quebrada San Julián was just one of many watersheds on the northern slope of the Sierra de Avila, in Vargas state north of Caracas, which had been affected. In the end, about 30,000 lay dead. They were killed by a series of debris flows. Part flood and part landslide, these debris flows—also named debris torrents—are fast-moving slurries consisting of storm water, streambed and bank materials like mud and sand and gravel, and even larger objects like boulders and uprooted trees.

Yet these watersheds don't contain huge, powerful, and notoriously flood-prone rivers like the Ganges or Mississippi. In fact,

7.1. Apartment buildings in Caraballeda destroyed by debris torrents on the Quebrada San Julián. Image courtesy of the U.S. Geological Survey.

at times the Quebrada San Julián can be nothing more than a low trickle of water. How did these nondescript little mountain creeks manage to wreak so much havoc? In this chapter, we'll take a look at the dangerous phenomenon of debris flows. And in doing so, we'll take a journey of exploration across some of the most exciting developments in mathematics, physics, and science as a whole over the last half century—in particular, the key ideas in complexity theory.

The archetypical debris flow proceeds something like this. Typically, under substantial heavy rain (and at certain times and places other hydrometeorological factors like rapid snow-melt or glacier recession), floodwaters race down steep gullies, tearing up whatever lies in their path and incorporating it into the main flow. Similarly, landslides occur on the adjacent steep hillslopes, dropping material into the creek below. The process can be self-reinforcing in several ways. Objects within the fast-moving flow slam into and dislodge more and bigger objects so that they too can be pulled into the torrent, acting in turn as still-larger projectiles to dislodge even bigger objects. The chunky soup of the debris torrent also provides so-called matrix support to larger objects. What does this mean? If you drop a rock into a slurry of moving gravel, the rock will have some trouble navigating its way past the gravel to the bottom: the gravel forms a matrix that supports the rock. The larger objects thus incorporated provide more support to yet larger objects. And by virtue of all the heavy material incorporated into the river flow, its overall density increases far beyond that of pure water, inducing buoyancy effects that make it easier for the debris flow to incorporate large objects. It's similar to how you float a little higher when you swim in the sea compared to a lake, because saltwater is denser than freshwater. That further increases the effective density of the flow and its ability to "float" even bigger objects. And as the banks are eroded by the torrent, the hillslope can further destabilize, causing yet more erosion. The net result of all these positive feedback loops is a channelized torrent of water-borne heavy projectiles, thundering down a steep valley at speeds of several meters per second, tossing around giant boulders and old-growth trees. It is vastly more potent than a water flood alone would be, granting even a small creek—under the right circumstances—the ability to easily destroy entire neighborhoods.

At a risk management level, the problem is exacerbated by the fact that as a debris flow runs out of steam at the base of its ravine, the materials it carries are deposited. Over time, and many such repeated events, this builds up into a relatively flat area somewhat like a floodplain. In the mountainous regions where all of this happens, and where convenient construction locations can be rare, people have historically tended to build their homes in these nice flat spots—right in the likely path of future debris torrents. This can be a problem all the way from coastal Venezuela to coastal California, the Italian Alps, and Japan. The most effective way to manage the risk is simply to avoid it altogether by not building in such areas. Where this is not a viable option, engineered solutions are available, though often not easy or cheap to implement effectively.

An important point about such slope failures—be they debris flows, snow avalanches, or landslides—is that the triggering event, such as prolonged heavy rain or an errant backcountry skier, is exactly that: only a trigger. This trigger is distinct from the underlying causes, such as geologically weak materials on steep slopes, poised to collapse. The same rainstorm a few months earlier or later might not trigger anything. Things have to be just so for the landslide to occur. Compare this to the operation of a rifle. It has to be loaded, the bolt has to be worked to put a live round in the chamber, the safety has to be disengaged—and then a light press on the trigger will set it off. But if any of these preconditions has not been satisfied, yanking on the trigger as hard and as often as you like is unlikely to do anything. There is, however, a crucial respect in which the analogy to a firearm fails. No third party is standing there on the mountainside, moving rocks around in just such a way that they'll be good to go for a tumble the next time it rains. It's as if the loaded and armed hunting rifle assembled itself from raw wood and steel and gunpowder. And this leads us to the much broader notion of self-organized complexity.

Self-organization and emergent behavior are concepts that arose in what some have called "postmodern" science to overcome limitations that more traditional methods had with explaining pattern formation. Traditional science is largely reductionist, meaning that the goal is more or less to split and simplify a question into more manageable bits, on the assumption that the answer to the big question can be had by combining, in some relatively straightforward and clear way, the solutions found for the smaller parts. This approach works extraordinarily well for many things, and in some sense has formed the basis of almost all scientific achievement and our modern technological society. It often does not, however, work so well for things like large-scale pattern formation in nature, if only for practical reasons of limitations upon computer speed or data. Perhaps the specific shape of that particular cloud forming overhead could in principle be explained from start to end by a mathematical-computational model based on classical physics, but it might require knowing the exact initial location and speed of every water molecule that came together to form the cloud and a computer run time longer than the age of solar system! Another, more philosophical issue with the classical view is that, in essence, it holds that complex patterns have no business forming spontaneously. This has to do with the second law of thermodynamics, which, stated simply, holds that disorder should increase over time. The second law of thermodynamics is right, but it's clearly not all there is to the problem: ecosystems, weather patterns, and the spiral arms of galaxies have all popped up seemingly of their own accord. In response to these challenges, a number of new theories and methods progressively developed over the latter half of the twentieth century, becoming fairly broadly established by about the 1990s or so. Often loosely collected under the rubric of complexity theory, the corresponding innovations included chaos theory (briefly described in a previous chapter), fractals (discussed later in this chapter), and cellular automata (to be addressed here in very short order), all of which initially were often presented, individually or collectively, as a revolution in science. These techniques have perhaps not quite lived up to such extravagant promises, in part because—at least from a practitioner's perspective—they often do not seem amenable to providing specific answers to specific questions: is that hillslope going to tumble down on my consulting client's house? What should I set as the sustainable harvest level for the sockeye salmon run on this river? Is it going to rain on my vacation next week? That apparent restriction aside, such views have proven remarkably useful for describing and explaining complex pattern formation in nature. A key point is that, in general, these fresh methods focus less on coming completely to terms with every little detail, and more on the big picture of how the system as a whole evolves.

Various different conceptual or mathematical devices have been used to this end. In cellular automata—as well as some other, similar approaches such as agent-based modeling and lattice models—the idea is to view the system's dynamics as a sort of accidental by-product of a great many very simple interactions between a great many very simple elements. Such models can produce extraordinarily complex behavior, little or no hint of which can be found in the model equations themselves. Such behavior is known as "emergent behavior," as it emerges from the model rather than being explicitly specified by the model. It is also an expression of self-organization, insofar as the model elements organize themselves or their collective behavior into patterns. And in self-organized criticality, the emergent behavior includes what can be viewed as system-wide "events" or "failures," not unlike the slope failures described earlier in this chapter. Indeed, it turns out that such cellular automata provide convincing explanations for, or at least very instructive ways of interpreting, many natural and social phenomena. And landslides are among them.

Credit for the first cellular automaton, or CA, seems to go to John von Neumann. The twentieth century saw a suite of astonishing scientific breakthroughs accomplished by unbelievably gifted people. Many were early- to midcentury central European émigrés to America, one of them being von Neumann, born Neumann János Lajos in Budapest. Most scientists would be happy to have a single research success sufficiently worthy to be noted outside industry reports, government documents, or refereed academic journals. Yet von Neumann, a mathematician, played major roles in everything from the Manhattan Project to the development of the modern programmable computer to game theory (an idea popularized in *A Beautiful Mind*, a Hollywood film starring Russell Crowe as John Nash, another major contributor to game theory). Von Neumann's ideas about cellular automata were motivated by an interest in the relationships between computing and biology, the notion being that both systems involve the processing of information (if this rings a bell, it's because we broached a similar idea in our chapter on information theory). In the end, he came up with a template for an artificial life form, the first cellular automaton. It involved a creature that lived on a grid, like a checkers or chess board, with each grid square amounting to a cell that could live, die, or reproduce according to various specified rules. A key point here is that the rules only directly govern the behavior of individual cells. There are no rules that dictate the behavior of the creature as a whole; it is the interactions between the individual cells, each responding automatically to certain circumstances in a little world immediately around it, that govern the overall behavior of the system. It's as if the things that you do as a human being—eating, sleeping, reproducing, dying, and so forth—are the net result of interactions between the cells in your body, rather than being governed by any overarching command-and-control system, and in fact there may be some truth to that! Others subsequently contributed much more to the field of cellular automata, often with an eye to exploring and explaining biological and ecological systems, and expanding the limits of computation. Particularly noteworthy for our purposes here, though, was the work of Per Bak, Chao Tang, and Kurt Wiesenfeld in the mid-1980s.

Their model is also known as the sandpile cellular automaton, because it loosely emulates the notion of placing grains of sand onto a sandpile. The mechanics of its implementation are pretty straightforward, and go like this. You have a grid. Within each grid cell, you have a random number of particles—grains of sand, say—between 0 and 3 of them. That's your initial setting. Then you move to the next time step. Note that in CAs like this, everything is quantized into discrete chunks: the size of the grid, the number of particles in each cell, time itself. And at the next time step (you could think of it being the next second, or the next day, or the next decade), a particle is added to a randomly selected cell, and then a redistribution rule is applied. The rule is as follows: define a four-point neighborhood, consisting of the grid cells immediately above, below, to the left of, and to the right of any given cell. If the cell you're looking at has 4 particles in it, they get moved to that surrounding neighborhood—each of the four surrounding cells receives one of the particles, and the central cell winds up with none. Taking $x$ to be the number of particles in a grid cell, and $i$ and $j$ to be the coordinates of that cell within the grid, this process can be symbolically represented as:

$$x_{i,j} \to 0$$
$$x_{i-1,j} \to x_{i-1,j} + 1$$
$$x_{i+1,j} \to x_{i+1,j} + 1$$
$$x_{i,j-1} \to x_{i,j-1} + 1$$
$$x_{i,j+1} \to x_{i,j+1} + 1$$

If, on the other hand, the cell you're looking at has fewer than 4 particles in it, do nothing to the particles in that cell. Then shift that redistribution rule across the grid, applying it to every cell, one at a time. For the cells lying along the edge (or corners) of the rectangular grid, the relevant neighborhood will include only three (edge) or two (corner) adjacent cells; the rule is applied exactly the same way, except that some of the particles being shifted out of the central cell will move not to an adjacent cell (because there isn't one) but instead will fall off the grid. Keep track of when that happens, and how many grains of sand are lost when it does. If there have been one or more redistributions during the time step, keep running this four-point operating rule across the grid until the redistributions stop—that is, until each cell in the entire grid has 3 or fewer particles in it. That's the end of the first time step. The second time step does exactly the same thing; it's kicked off by adding a particle to a randomly selected cell in the grid, and then the redistribution rule is applied in exactly the same way as before. This simple process continues, time step after time step. Sometimes, no cell has more than 3 particles in it after the "kick-off" particle is randomly added at the start of the time step, so nothing really happens. In some other time steps, one redistribution will lead to another redistribution and so forth—a small example can be seen in figure 7.2—creating a cascade or "avalanche" of particles off of the grid. We can see that the sandpile CA really is like a pile of sand in a basic, intuitive way: add one grain to the top, perhaps nothing happens; add another, and maybe then the slope is too steep so it rolls down, dislodging a few others, which in turn dislodge more, creating a little landslide. That's all there is to it.

Take a look at the figure to get a better sense of how the process works. Panel (a) gives the starting point, or "initial condition," with a random number of particles in each cell. Cells are identified by their horizontal location, $i$, and vertical location, $j$, as shown. Panel (b) illustrates the first time step: one particle is added to a cell chosen at random—in this case, cell (2, 3). We then check to see if any cell has more than 3 particles in it. None does, so the time step is over: no landslide occurs. Panel (c) gives the second time step: a single particle is again added to a randomly selected cell—here, cell (1, 1). We then check to see if any cell has 4 particles in it. Cell (1, 1) does, triggering an avalanche. Panel (d) shows a continuation of this second time step. We follow the redistribution rule: one of the four

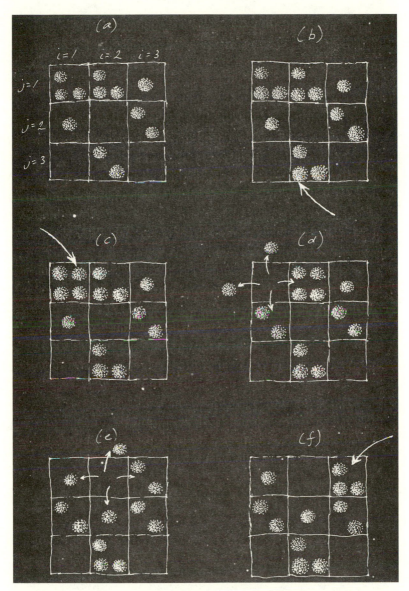

7.2. A few time steps in the evolution of a very simple (3 x 3) sandpile CA. In practice, much larger grid sizes are used, with computer implementation.

particles in cell $(1, 1)$ is moved to each of the four adjacent cells, $(i + 1, j) = (1 + 1, 1) = (2, 1)$; $(i − 1, j) = (1 − 1, 1) = (0, 1)$; $(i, j + 1) = (1, 1 + 1) = (1, 2)$; and $(i, j − 1) = (1, 1 − 1) = (1, 0)$. Of course, there is no row $j = 0$ or column $i = 0$, so that means that two particles are lost—they're material sliding off the grid. The second time step extends into panel (e), where we see the landslide continuing to cascade through the grid. Specifically, we check again to see if any cell has 4 particles in it, and now cell $(2, 1)$ does. So its contents get moved to adjacent cells, leading to the loss of one more particle from the grid. We then do a final check and see that, following this second redistribution, no cell has more than 3 sand grains in it. The landslide is over, involving a total loss of 3 particles from the grid. We move on to the third time step in panel (f): one particle is again added to a randomly selected box, in this case cell $(3, 1)$. We check to see if any box has more than 3 grains in it. None does, so the time step is over—no landslide occurs at this time.

How could anything interesting or relevant be produced by such a simple and seemingly arbitrary set of rules? Well, remember that emergent behavior is all about generating complex dynamics and patterns from simple interactions— and it turns out that the sandpile cellular automaton is a poster child for emergent behavior. The rules only explicitly involve a maximum of five grid cells at a time: the cell it's being applied to, plus its neighborhood (an additional two cells if the cell is at a corner of the grid, three cells for an edge, or four cells in the interior of the grid). And only a single particle can be added to the grid at a time. Yet avalanches occur that involve many particles, over the whole grid. That is, catastrophic events are generated, much larger than anything explicitly indicated in the system's laws. This is called self-organized criticality: a system criticality or failure or event that comes about from the way the system organizes itself, rather than being overtly specified in an overarching and specific law or rule or equation. And what's more, it turns out that the self-organized criticality of the sandpile CA shows fractal dynamics.

Fractals per se were only recently discovered—the key work was done by Benoit Mandelbrot in the second half of the twentieth century—yet fractals have turned out to be one of the most ubiquitous phenomena in the universe. Fractals involve an infinite progression of patterns lying within patterns within patterns, like looking at someone painting a painting of you looking at someone painting a painting of you looking at someone painting a painting of you, each set of repeating patterns being the same (or at least similar in character) except for its size. Such fractal geometries have been identified in everything from the distribution of galaxies in the night sky, to the way that twigs split off from tree branches that split off from larger branches and so forth. Fractals are ubiquitous in the geophysical sciences as well. One example is the drainage pattern of a river system. Typically, many tiny streams feed into somewhat larger and fewer streams, which in turn feed into still larger and fewer tributaries to the main river. This drainage pattern is called dendritic. River systems ranging in size from a few-millimeter-wide trickle through your backyard in a rainstorm, to mighty continent-spanning drainage basins, all tend to show a similar overall pattern. And so, looking at maps or aerial photos of watersheds, they all look in some sense pretty much the same, and we need something in the photo to give us a visual clue of the scale—such as a scale bar, of course.

Fractals also occur in time. The interpretation is analogous to that of geometrical fractals. Small frequent variations in river flow (for example) are added onto ever-so-slightly larger and slower events, which are in turn superposed on top of still bigger, longer term variability, and so on. Consider the two plots of river depth (vertical axis) as a function of time (horizontal axis) shown in figure 7.3. I've deliberately omitted tick marks and labels in these graphs. Without those cues, can

**(a)**                           **(b)**

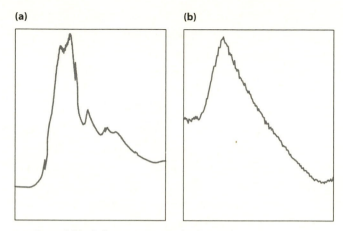

7.3. Fractal-like behavior in storm hydrographs.

you tell what the time frame is for either of these plots? Is it months—the waxing and waning of the wet season, say—or is it just a few hours under a passing storm front? And can you tell how much the river is rising and falling over that time period? Is it 10 meters or is it 10 centimeters? In fact, these images are both for the same creek during the same storm. On the left we see depth variations over a five-day window. The image on the right is in fact a blow-up of one of the small "blips" in the plot on the left, showing depth variations over just fifteen hours. The vertical scale on the left spans about 2 meters, whereas the one on the right spans only 35 centimeters. Obviously the two graphs are not identical, but the overall pattern is similar, with a flat initial part, a steep rise, a sharp peak, and a gradual decline back to another nearly flat part. This similarity of form over many different timescales is a feature of fractal dynamics, in which a small, short-lived event is superposed on a larger, longer-lived event, and that upon a larger and slower one and so forth. In this particular example, we're seeing the effects of episodic rain showers within a larger storm system.

In a so-called dynamical or temporal fractal such as this, there isn't just one characteristic timescale. The resonance frequency at which a church bell rings, or a guitar string vibrates, are examples of a single characteristic timescale. But in dynamical fractals, the thing of interest shows variations across a wide range of timescales, all similar in form but different in size and frequency. Examples of fractal time series include river flow (as noted above), stock price volatility, black hole x-ray emissions, background noise in electronic devices, global temperatures, and human cardiac activity (it turns out that a fractal heart is a healthy heart, by the way). It also turns out that fractal dynamics are in turn closely related to or synonymous with a number of other phenomena. One is the Hurst effect, named after the physicist-turned-hydrologist Harold Edwin Hurst, who discovered it while working on preparations for the design of the Aswan Dam in Egypt. The effect was uncovered in an amazing six-century-long record of yearly minimum water elevation of the Nile River near Cairo. This so-called Nilometer time series, introduced in chapter three, is one of the more notable data sets in all of science, as it provided the prototype of fractal dynamics and the Hurst effect. Another very close relative of dynamical fractals is a phenomenon that had long been observed in many types of data, and which has been called by various names like power-law spectral scaling, flicker noise, $1/f$ noise, pink noise, and a scale-free spectrum. This is a special case of the red noise we described in our earlier chapter on spectral analysis. And as you might also recall from that chapter, red noise is in turn related to the memory of a system. In a $1/f$ process—a name that derives from the proportionality of the power spectrum to the inverse of frequency—that memory is, at least in principle, infinitely long.

There are a number of consequences to such dynamics. Extreme events like floods (or stock market crashes!) cluster together in time, which can make their consequences direr, as you're getting hit with one disaster right after another, without time to recuperate in between. And these extremes may also be more severe—that is, both higher highs, like worse floods, and lower lows, like worse droughts—than one would be led to believe using conventional statistical methods and assumptions. Additionally, long upward or downward trends are more likely to occur by chance in a fractal process than in white noise, yet can disappear as abruptly as they appeared. That has the potential to complicate analyses of historical data records for climate change effects on air temperatures or river flows, for example.

Dynamical fractals also present a more fundamental scientific challenge: to understand how the statistical physics of fractals relates to the deterministic physics of underlying processes. This is something that scientists generally haven't done such a great job at so far. In hydrology, it's likely that fractal dynamics reflect a combination of physical causes. As we discussed in our previous chapters on streamflow memory and spectral analysis, red noise in streamflow is the signature of memory-inducing water storage mechanisms in a watershed, ranging from tree canopies to soils, lakes, wetlands, and aquifers. In the case of fractal river flows, there are many different water storage mechanisms within the upstream watershed, each with a relatively simple but different timescale of response to weather inputs, and which in aggregate sort of "smear" together downstream to create a continuous $1/f$ red-noise spectrum. Additional contributions to fractal dynamics in streamflow may come from atmospheric drivers of hydrologic variability. For example, some work suggests that rainfall is a dynamical fractal as well. And certain models of the physics of ocean temperature suggest the occurrence of $1/f$ scaling there; it may then propagate into long-term river flow variations because ocean temperature is a regulator of regional and global climate. There is still much, however, to be discovered about fractal dynamics in rivers.

Another very closely related expression of fractal dynamics is in terms of inverse frequency-magnitude relationships. We're all familiar with the basic notion behind this, just from everyday experience. Consider car accidents. Virtually all of us have been in a minor fender bender. Fewer have been involved in more serious accidents where someone was injured. And thankfully, only a relatively small proportion of us have experienced major, perhaps crippling or fatal, car crashes. That is, there is an inverse relationship between frequency and magnitude: small accidents are common, medium-sized accidents are less frequent, and large accidents are comparatively rare. We see this general sort of relationship in many aspects of life and the world around us. To qualify specifically as a fractal process, more stringent requirements are imposed—in particular, a power-law relationship between frequency and magnitude is key. What this means is that if we draw a graph of how often events of a given size occur, for a whole range of sizes, then that plot is a straight line on double-logarithmic graph paper. We discussed double-logarithmic plots before: to recap, it's just a way of graphing results so that each major tick mark on the graph corresponds to the next multiple of 10, like the sequence 0.01, 0.1, 1, 10, 100, and 1,000. An example is given in figure 7.4.

And although this is a pretty specific type of relationship between frequency and magnitude, it turns out that an amazing range of dynamical systems obey it. Earthquakes are one example. In this case, the power-law association is known as the Gutenberg-Richter relation: there are many small magnitude 2 tremors, but only a handful of city- or nation-destroying magnitude 8 megaquakes. The relationship between the size of volcanic eruptions and how often eruptions occur is another example. And asteroid impacts. And forest fires, floods, acid rain deposition, species extinctions, and even wars for that matter! And, as it turns out, landslides and debris flows.

Which brings us back to our sandpile cellular automaton. In fact, the above figure of a power-law relationship is the output from such a CA. To be sure, it's much larger than the cellular automaton illustrated earlier in this chapter, having hundreds

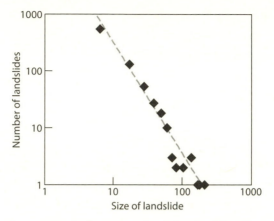

7.4. Power-law magnitude-frequency relationship for debris flows, as generated from a sandpile CA. The horizontal axis gives the number of particles involved in a cascade, and the vertical axis gives how many cascades of that size occurred.

of cells on a side instead of just three, and it's been run much longer, with careful tallies taken of how many avalanches of what size happened. But the simple, local rules by which its evolution are governed are exactly the same. Von Neumann's artificial life has come a long way. It may also have a long way to go yet. Certainly, the detailed physical dynamics of landslides are far more complex than what is represented in a sandpile cellular automation, and far more involved CA models—not to mention other types of mathematical landslide and debris flow models—have been developed since the largely theoretical work of physicists like Per Bak and his colleagues several decades ago. Yet it remains remarkable that such a simple model, involving little more than one idealized grain of sand falling upon a few others, can generate such complex (and realistic) behavior as self-organized criticality and the fractal dynamics of landslide activity. It is perhaps still more remarkable how widespread, indeed almost universal, such forms of complexity are throughout the geophysical sciences and far beyond.

# 8

## THE SKY'S NOT THE LIMIT

To understand the river at your feet, sometimes you have to look to the stars above. In this chapter, we'll explore linkages between rivers and a few considerations beyond Earth's atmosphere: astronomical forcing of climate, glaciers, and water (that is, the fact that what happens in space actually impacts what happens in watersheds) as well as the use of satellite remote sensing in water resource science. Along the way, we'll again come across a number of surprising facts and linkages: river flows depend on how the earth's orbit wobbles and the sun's output fluctuates, grand Renaissance paintings of wintertime river scenes illustrate genuine water management issues that societies of the past had to cope with, the continued waning of the last ice age may today affect the futures of billions of people in China and India, and espionage and watershed mapping have something in common.

Consider the two largest and simplest forms of variation in river flows. Both are ultimately astronomical in origin, yet they're so commonplace, so much a part of our normal experience, that we may fail to think of them as being otherworldly. I'm referring, of course, to the days and seasons. Let's start by revisiting the origins of these familiar variations. Earth spins on its rotational axis once every twenty-four hours, defining the day, and orbits around the sun once every 365.25 days, defining

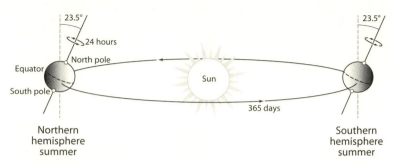

8.1. Schematic illustration of Earth's orbit.

the year. Seasons come about mainly from the interaction of this yearly orbit with a tilt in Earth's rotational axis. Defining the plane of Earth's passage around the sun as "horizontal" as shown in figure 8.1, the axis of Earth's daily rotation isn't straight up and down, but instead inclined relative to the vertical at an angle of 23.5°. It always points to about the same point in the distant night sky—toward the North Star, also called Polaris, providing a traditional navigational aid. At what's shown here as the left-hand side of Earth's yearly orbit around the sun, the sun's rays impinge more directly on the northern hemisphere than on the southern hemisphere. That is, the daytime sun is more "overhead," so to speak, providing a greater solar energy flux and creating the northern hemisphere summer. Half a year later, Earth has orbited to the opposite side of the sun, but its axis is still pointed in the same direction, toward Polaris. So instead, more of the southern hemisphere is now directly facing the sun during the day, creating the southern hemisphere summer.

We've already discussed seasonal cycles several times, as there is little if anything in natural (and even most human-altered) watersheds that affects rivers as dramatically as the passage of the seasons. The daily waxing and waning of sunlight as Earth spins on its axis, and the corresponding rise and fall of temperature on an hour-to-hour basis, is a less obvious, yet also very important, factor in terrestrial hydrology. One example is day-night variation in plant water uptake. Compare the lush greenness of a rain forest to the sparse vegetation of a desert, or look at the vast amounts of water needed for agricultural irrigation; water is crucial to plants. Loosely speaking, pores in the leaves, called stomata, open up during the day to let in atmospheric carbon dioxide—what plants "breathe" as they manufacture their own food by photosynthesis. When open, though, these stomata also allow water vapor to escape from the plant; this is the process of transpiration. That escape of water vapor results in a loss of water from the plant during daytime. It also sets up a pressure gradient between the leaves and the base of the plant—forming a sort of pump, allowing the roots to take up water from the soil and then facilitating its upward transport (through a tree trunk, for example). Plants are thrifty in the sense that they conserve water by closing their stomata and shutting down the whole process at night, when the sunlight needed for photosynthesis isn't available anyway. The overall impact of transpiration on the water budget of a river can be huge, and a corresponding diurnal variation can be directly seen in some measurements of soil moisture and streamflow. Another example is the dramatic daily rise and fall of water levels in rivers fed by the melt of snow and glaciers. Flows increase after the heat of the day, and then gradually drop as nighttime cooling shuts down snow and ice melt, only to start up again the next afternoon.

The moon provides another form of astronomical forcing with a part to play in hydrology. The moon, the ocean tides exerted by its gravitational pull on our spinning water planet, and the phases it passes through during its twenty-seven-day orbit around Earth, have deep implications for life—particularly noticeable in various biological rhythms—and thus also permeate global culture. Ancient lunar or lunisolar calendars, based in part on the moon's phases, may have been both an anthropological reflection of the moon's significance to life and a capitalization on the regularity of the changing moon as a time-keeping device. These are still used today for religious purposes across an astonishing diversity of faith groups; examples include the Theravāda Buddhist calendar, the Zoroastrian calendar, and the *computus*, an algorithm for calculating the date of Easter that attracted the attention of none other than Carl Gauss, one of history's greatest mathematicians and a progenitor of the statistical "bell curve," also known as the normal or Gaussian distribution. The moon's role as a logistical factor in the broader narrative of human events is also unmistakable. In the age of sail, ships had to wait for a favorable tide to sweep them out of port. The full moon of autumn signals the harvest and the hunt, rising shortly after sunset for several days in a row and so effectively extending the day with its powerful reddish light when it's most needed. The date of the Allied invasion of Normandy during World War II—the largest seaborne invasion in history, better known as D-Day—was selected in large part for its full moon and a rising tide, illuminating the nighttime beaches and optimizing tactical conditions for the amphibious assault. Perhaps the most singular episode in human history, a grand intersection of symbolism and technical achievement, was the landing of man on the moon. And as for rivers? The moon—in particular, the tides—also have a strong bearing on hydrology. The associated gravitational variations ever-so-slightly squeeze and then release soil and rock as if it was a sponge. These so-called Earth tides cause water levels in some artesian aquifers to exhibit cyclical variations. Additionally, recent research suggests that for watersheds having an unusually strong groundwater contribution to flow, these effects can in turn be observed in river water levels. The most notable ramification, though, lies with coastal flooding. In the largest such floods, three key ingredients are often involved: rising river levels, reflecting rainfall from a storm system; a temporary local rise in sea level called a storm surge, caused by a complicated mixture of wind effects and low atmospheric pressure from that same storm; and a high tide. This may seem like an unlikely confluence of events, but the first two ingredients can often go hand in hand and the third is commonplace. Together, they conspire to bring about some of the worst flooding events in coastal communities, which are often port cities and general transportation hubs that are built on the flat floodplains of major rivers where they meet the sea. More generally, ocean tides can be felt in both aquifer and river water levels some distance inland from the sea, and the lowermost portions of a river can sometimes even flow "uphill" toward the headwaters for a short period of time when ocean water levels are high. Earlier, we offered the Thames as an example of such effects. Associated flooding of London was a motivation for the massive gates of the Thames Barrier.

Moving to longer timescales, few natural phenomena capture the public imagination quite like the ice age does. Just think of the movies made about it, ranging from the Disney cartoon franchise of the same name, to the climate change disaster movie *The Day after Tomorrow*. Perhaps it has this impact because it lies just on the edge of modern civilization's living memory. It ended roughly 12,000 years ago, and archaeological discovery extends the history of contemporary societies that far back in a more or less unbroken line. In some sense, the ice age effectively bridges the deep time of geologists and the human time of anthropologists. So, standing in New York's Central Park today, looking down at the grooves (called striations) left in the bedrock from when the last ice sheet slid over it, and then looking up at the soaring Manhattan skyline, it's pretty easy to visualize how the Upper West Side might have been occupied by a towering glacier. In truth, though, a whole series of ice ages came and went over the two million years or so of the so-called Pleistocene geological epoch. An astronomical theory for the origin of the ice ages emerged in the early twentieth century, put forward by the geophysicist Milutin Milanković, after which it is named. Firm evidence for this orbital theory appeared about another half century later. It was provided by spectral analysis of paleoclimate data, which were obtained through careful study of core samples taken from sediment drilling in the deep sea—in particular, depth-dependent variations in the isotopic composition of the buried skeletons of tiny sea creatures captured in that geologic record. Many other factors were also at play in generating the growth and retreat of the Pleistocene ice sheets, but Milanković cycles are the central organizing theme for our understanding of the ice ages.

These cycles consist of variations in Earth's orbit. Recall that the axis of rotation on which Earth spins on a daily basis is inclined by 23.5°. It turns out, though, that this is just the current value. The number actually swings from 22.1° to 24.5° and then back again, completing a full cycle over about 41,000 years. What's more, this axis of rotation also wobbles around, a little like how a spinning top begins to wobble as it loses speed. This is called precession, and a full wobble takes approximately 23,000 years. The third major Milanković cycle involves the Earth's orbit around the sun, which is not a perfect circle, but an ellipse. The amount of deviation from a true circle, called the eccentricity of Earth's orbit, oscillates with a period of about 100,000 years. Together, these three astronomical phenomena deeply affect the seasonal distribution of heat energy that Earth absorbs from the sun, in a way that physically explains the ice ages and which has been confirmed and reconfirmed in paleoclimate studies.

How does the ice age connect to rivers? In a previous chapter, we talked about how the carving action of moving glaciers during the last ice age determined where many rivers flow. Perhaps an even more important implication lies with the fact that the ice sheets and mountain glaciers of today—which collectively form more than half of the planetary freshwater volume—are the remnants of the last Pleistocene ice sheets. It's hard to overestimate the significance of this simple fact. In particular, the ice fields of most major mountain chains serve as what have been widely called the world's "water towers." Ice and snow melt from the peaks of the Rockies, Alps, and Andes, for example, feed the major river systems of North America, Europe, and South America. This gift, left for us by the last glaciation, helps provide water for literally billions of people living downstream. And the continued waning of the most recent glaciation—the last, fading modern artifacts of past Milanković cycles—therefore have sharp repercussions for our present and future water resources.

Let's focus on the Hindu Kush, Karakoram, and Himalayan Mountains and their glaciers. First, some background. This series of overlapping mountain ranges wraps around the northern edge of the Indian subcontinent, a result of the ongoing tectonic collision between the Indo-Australian plate and the Eurasian plate, creating the tallest peaks on the planet and lifting up the Tibetan Plateau. It's due to such great altitude that these mountains, lying at subtropical latitudes, continue to host one of the world's largest remaining icefields—and greatest freshwater sources. In fact, the Himalayas and adjacent mountain ranges are the source of the Indus, Yangtze, Yellow, Ganges, and Brahmaputra Rivers among others, and thus help provide water to a full quarter of the global human population.

We have to pause for a moment here to explain the monsoon, because a considerable portion of the water volume in many of these rivers originates, in fact, as rain during the Indian monsoon season. A monsoon is a seasonally reversing wind system that sets up rainy and dry seasons. Water has a greater capacity to absorb heat energy without actually changing temperature than rock does—that is, it has what's called a greater specific heat capacity—which is why coastal locations often have a seasonally more stable climate than continental interiors. As a result, the Indian land mass heats up more than the surrounding Indian Ocean during summer, and the warm air rises convectively. Something has to fill up that void, and that "something" is moist air from the Indian Ocean. This sets up a wind pattern from the ocean to the land. As the moist ocean air subsequently heats up over the land, it also rises convectively, and then cools adiabatically—we discussed this in our chapter on information theory—and rain is generated. A lot of rain, which is why the word "monsoon" is synonymous with "deluge." During cooler months, the pattern reverses, with the continent cooling off more than the ocean, such that the air over the Indian land mass sinks and then flows out to the Indian Ocean.

But in addition to these monsoon contributions, the streamflow signal from snow and ice melt in the mountains is also unmistakable, forming a critical baseline flow level for the major river systems of India, China, and other Asian nations. Consider the Indus River. With headwaters in the soaring peaks of the border region between China, India, and Pakistan, it runs across the semiarid plains of Pakistan to drain into the Arabian Sea. Indus water levels typically start rising in spring from prodigious snow and ice melt in the mountains, followed by monsoon rains. The water resources provided by the Indus are in turn critical to Pakistan's agricultural production and economy.

More broadly, alpine glaciers provide a key additional water source in any river basin they occupy, and in a world with growing populations and economies, that's important. A streamflow effect appears with even as little as 2% glacier cover in the upstream watershed, and those contributions can be most pronounced when they're needed most: during warm, dry periods, when glacial ice melts and other runoff sources aren't available. Further, the abundant cool water of glacial rivers has profound implications for aquatic species diversity and fish populations. And glaciers also affect the way that rivers respond to climate variations. Two adjacent rivers can exhibit completely different reactions to phenomena like El Niño, for example, if one has an upstream glacier available for melting and the other doesn't.

If the critical water sources of the Himalayan ice fields and other alpine glaciers worldwide are the final vestiges of the last ice age, then what will the future bring? Overall, the trend seems to be toward an accelerating loss of glacial ice and, over the long term, the associated water resources. Contemplate the implications for the Ganges River of India and Bangladesh. The most sacred river in Hinduism, embodied in the goddess Ganga, the devout travel to its shores in life to be cleansed of their sins, and in death to be swept to salvation. The river flows roughly southward from its source on the Gangotri Glacier in the Himalayas near the Chinese border. Emerging from the mountains, the Ganges turns east near the ancient pilgrimage city of Haridwar and runs across the great Indo-Gangetic Plain, a breadbasket of India. After crossing into Bangladesh, the Ganges merges with another of the world's great rivers, the Brahmaputra, to form the planet's largest delta. But the glaciers in the Himalayan headwaters of this great river—an important contributor to its flow regime—have been receding rapidly, as illustrated in the satellite image of figure 8.2. The hydrologic fundamentals of the Ganges River are changing. Contemporary shifts in mountain glaciers and their hydrologic output are key water resource questions going forward, not only for the Ganges, but globally.

These changes have, in fact, a variety of sources. We mentioned above that these glaciers are in part the fading artifacts of a long-gone climatic era whose passage was driven by very long-term orbital cycles. We also discussed global anthropogenic climate change in a previous chapter. For most mountain glaciers, the implication of human-caused climate change is a loss of glacier ice due to increasing summertime melting under warmer temperatures. That can initially mean greater flows in

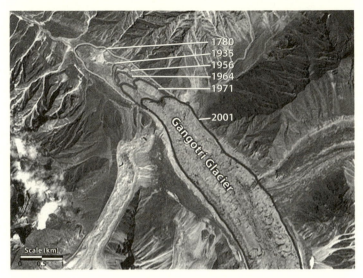

**8.2.** This NASA image uses reconstructions, and high-resolution satellite images from the Advanced Spaceborne Thermal Emission and Reflection Radiometer, to illustrate recession of a large glacier in the headwaters of the Ganges. Image by Jesse Allen, Earth Observatory; based on data provided by the ASTER Science Team; glacier retreat boundaries courtesy the Land Processes Distributed Active Archive Center.

some glacier-fed rivers as the taps are turned on, so to speak, but eventually the effects of progressively smaller glacier area take over, resulting in a streamflow decline. Some projections call for dramatic glacial recession, and the eventual obliteration of glaciers altogether from certain regions—along with the seasonal meltwater pulse they used to supply. Deposition of airborne contaminants, in particular industrial soot, on top of the ice may also be an issue. Dark particulate matter like soot decreases the albedo of ice—a concept we raised in a discussion about positive feedback loops in our chapter on why rivers are where they are. Just like when you get a lot warmer if you wear a black shirt outdoors on a sunny day than if you'd worn a white shirt, a coat of dark soot makes the ice more susceptible to melting. That process could reinforce both natural and anticipated human-caused climate change impacts. And all of this is happening against a backdrop of ongoing glacier recession following the end of a period called the Little Ice Age.

This brings us to another form of hydrology-relevant astronomical climate forcing. Earth's orbital parameters are variable, as we discussed above—and so too is the sun's output. Sunspot counts provide a useful measure of variations in solar energy. These dark spots on the sun's surface are cooler than their immediate surroundings but are associated with an overall increase in solar activity. The more sunspots, the more energy released from the sun's surface and sent on to Earth. As a way of monitoring solar activity, sunspots have the advantage that they are relatively easy to observe (this of course needs to be done carefully so as to avoid injuring your eyes), which means that very long records of the number of sunspots have been kept. Sunspots, and the associated solar energy, exhibit an eleven-year cycle. Peaks in that cycle correspond to increased solar storms, and this "space weather" has nasty effects on modern technology: frying communications satellites, threatening the safety of astronauts, and shorting out earthbound power grids, causing blackouts. In fact, both NASA and the US National Oceanic and Atmospheric Administration have launched satellites into solar orbits to provide a distant early warning of massive solar storms like coronal mass ejections, and the US National Weather Service even operates a space weather prediction center. Evidence for impacts of this eleven-year cycle specifically in climate and hydrology, on the other hand, is spotty—some analyses suggest it, others don't. A more pertinent feature of the sunspot record is a long period of generally low sunspot, and therefore low solar, activity spanning about AD 1280 to AD 1820. This series of sunspot minima, which were separated by brief recoveries, included the so-called Wolf, Spörer, and Maunder Minima. The low solar activity generally corresponded to a period of cold, wet climate known as the

Little Ice Age. This was in turn preceded by a period of elevated sunspot counts starting around AD 1100 called the Medieval Maximum, which loosely corresponded to a climatic phase named the Medieval Warm Period. Additionally, the Little Ice Age was followed by another period of elevated sunspot counts starting around AD 1900 called the Modern Maximum, corresponding to today's warmer climate. The exact physical mechanisms whereby moderate fluctuations in energy received from the sun generate large climatic responses remain incompletely determined, and one must always bear in mind that correlation is not causation. On the other hand, our ability to fully understand something is not a prerequisite for that thing's existence, and most scientists agree there is some kind of real link, possibly involving a solar radiation-induced modulation of atmospheric chemistry. Likewise, research is ongoing as to just how global in scope these past climatic periods actually were, but it's clear that the effects of the Little Ice Age were very widespread, and the Medieval Warm Period was certainly an important fact of life in the North Atlantic region.

Regardless of the details, both the Medieval Warm Period and the Little Ice Age had major implications for world history. As a result, they also continue to offer lessons for us about today's shrinking glaciers and, more broadly, about how societies may respond to climatic shifts. It's now well established that Columbus was not, in fact, the first European to set foot in the Americas. That distinction goes to the Vikings. Best known for their raids, they were also traders, explorers, and settlers. They hopscotched across the islands ringing the North Atlantic, establishing towns and farms in Iceland, Greenland, and finally, around AD 1000, Vinland—what's known today as Newfoundland. But while Christopher Columbus's arrival in the balmy Caribbean signaled the start of the largest wave of conquest and immigration the world has seen, Leif Erikson's Viking experiment sputtered and eventually fizzled out. Why?

There were undoubtedly many reasons, but a key factor was a changing climate and its implications in the far north. The Vikings set out during the Medieval Warm Period. When they arrived in Greenland, they called it that because it was green. But after a while, it stopped being green. Their farms failed; they retreated. That transition to a colder, wetter climate marked the start of the Little Ice Age, which began about AD 1300, picked up speed around AD 1500, and lasted right through to the mid- to late nineteenth century. The glaciers of the Alps advanced, erasing Swiss mountain villages. The great rivers of western Europe, like London's Thames and Paris's Seine, which rarely if ever freeze today, started freezing over in winter on a regular basis. These effects were widely recorded in paintings of the time (fig. 8.3). One important hydrological implication is that today's widespread retreat—in the Himalayas and elsewhere—of alpine glaciers, which can exhibit drawn-out responses to climatic forcing, may in part be a continued recovery from the frigid conditions of the Little Ice Age. Other effects, such as global anthropogenic climate change, are added on top of that.

A particularly intriguing implication of the Medieval Warm Period and Little Ice Age lies with the evolution of water management and engineering strategies. Today, Holland is home to the world's master hydraulic engineers. But that expertise didn't just appear overnight, and the Dutch experience provides an opportunity to explore how a technologically advanced and strongly organized culture dealt with well-documented hydroclimatic shifts of the past. The results may be a little surprising. The Netherlands are quite flat, and thus exposed to both the river floods of the Rhine Delta (actually an interfingering set of deltas from several rivers including the Rhine) and storm surges from the North Sea. Up until the start of the Medieval Warm Period in the tenth century, the small population of the Netherlands lived mainly on higher ground. The drying of climate around AD 950 prompted the Dutch to expand

8.3. *Jagers in de Sneeuw* (Hunters in the Snow), painted by Flemish artist Pieter Bruegel the Elder in 1565. Note the ice skaters on the frozen river in the background. Many such paintings appeared around this time, illustrating harsh climatic conditions in Europe. Image courtesy Wikimedia Commons / Public Domain.

into wetland areas that were previously difficult to inhabit or farm, but that shift still required a little technological help, and the ingenious Dutch proceeded to implement new drainage techniques. Unfortunately, the act of draining these wetland areas led to land subsidence, not unlike that occurring today in California's Central Valley under groundwater mining, as we discussed in a previous chapter. Ironically, this lowering of the land surface led to an increase in flooding. This was exacerbated by large-scale deforestation in the upstream German parts of the Rhine, which can reduce a basin's ability to delay storm waters. In response, the Dutch began work on their tremendous system of dikes, the construction of which helped prompt a centralization of government authority. By 1350, all the major rivers of the Rhine Delta were fully diked.

A key turning point was the advent of the Little Ice Age in the late 1400s or so, introducing a brand new type of flood risk to Dutch rivers: ice-jam flooding. Today mainly a feature of northern rivers in places like Canada and Russia, ice-jam floods occur when a river's surface freezes, but then breaks up into a jumble of ice blocks that temporarily dam the water flow. To make matters worse, the dikes had unintended consequences: no longer able to spill their banks and deposit their sediments on the floodplains, rivers began silting up their own channel beds and gaining elevation, making it more likely that they would overtop those dikes. The introduction of ever-more-sophisticated technologies for drainage, such as the iconic Dutch windmills that powered the dewatering systems, helped over the short term but exacerbated the problem over the long term by accelerating subsidence of floodplains adjacent to the rivers and dikes. By the late eighteenth and early nineteenth centuries, serious river floods were commonplace and a major issue in the political agenda of the Netherlands. However, the late 1800s and early 1900s saw a tremendous reduction in river flood risks, under massive structural improvements to the water management systems and an amelioration of climate as the Little Ice Age drew to a close. Today, flood risks seem to be on the rise again, perhaps due to contemporary global climate changes. The Dutch continue to adapt, implementing their Room for Rivers program, which restores some natural functionality to rivers—including flood mitigation—by setting aside riparian areas. The overall lesson seems to be that the choices we make about watershed management and water resource engineering, both politically and technically, can either mitigate or intensify the impacts of climatic shifts.

So far, we've spoken mainly about how astronomical forcing can affect rivers. Yet there is another, very different, aspect of the relationship that outer space has with hydrology: satellite-based Earth observation systems, and how they provide crucial data on weather, water, and land surface characteristics important to hydrologic analysis, modeling, and prediction. One example is the Gravity Recovery And Climate Experiment (GRACE) mission. Recall, from our chapter about buried treasure, that careful mapping of tiny variations in Earth's gravitational field allows us to infer the geology below. Different types of Earth materials have slightly different densities and therefore slightly different contributions to the total gravitational pull. GRACE extends this concept. The mission consists of a pair of satellites, and small variations in their orbits give information about the gravitational field. This in turn provides valuable data about large-scale variations in the water budget of entire basins.

Most satellite remote sensing, though, involves spaceborne instruments that monitor particular types of electromagnetic radiation—in other words, light. Visible light consists of electromagnetic (EM) waves having frequencies that our eyes can detect. Other parts of the EM spectrum can only be monitored using specialized equipment. For example, infrared radiation, which you can think of as heat, has a lower frequency than visible light and can be detected using night-vision goggles employed by military special forces. Conversely, the x-rays used to image the bones in your arm after your mountain biking accident correspond to higher frequencies than the human eye can see. Some satellites have instruments that monitor EM energy passively. Essentially, they take photographs. Others send out a signal and record the response—like radar, which operates in the microwave range, at even lower frequencies than infrared. And some satellites conduct this EM imaging as they move in low, fast orbits, whereas others move in high, slow orbits.

What information a satellite can provide depends on the EM bands it monitors, whether it has active or passive sensors, and the nature of its orbit. For instance, infrared sensors allow the satellite to take images of clouds at night, when a sensor operating in the visible spectrum (a normal camera) would show only darkness. Radar extends this further: it can see the surface of Earth through the clouds, day or night, as if they weren't even there. There are also trade-offs. Consider the orbital characteristics of a satellite moving across the skies overhead. Satellites are never just stationary in space; they would simply fall back to the ground. Rather, an object in orbit experiences a delicate balance between its velocity as it whips around Earth (which wants to send it flying out into space) and Earth's gravitational tug on it (which wants to crash it back down into the atmosphere). At greater altitude, Earth's gravitational pull is weaker, so less speed is required to keep the satellite aloft. Of particular note is the altitude of 42,164 kilometers above the center of the planet, or about 36,000 kilometers (roughly 22,000 miles) above Earth's surface. The spacecraft speed needed to precisely balance Earth's gravitational pull at that altitude happens to correspond to a satellite orbital period around the Earth of exactly twenty-four hours—which of course is also how long it takes Earth to spin through a day. This is called a high Earth, or geosynchronous, orbit. If the satellite's path also happens to be located directly above Earth's equator, then the satellite follows Earth as it spins, maintaining a fixed position in the sky above (relative to the ground). By hovering over a particular location, such so-called geostationary orbits allow the same area to be monitored twenty-four hours a day, seven days a week. They're used, for example, by communications satellites (which is why your satellite TV antenna can remain pointed at a particular spot in the sky, instead of having to actively track a satellite) and weather satellites (such as the Geostationary Operational Environmental Satellites, or GOES, which form a backbone of operational weather forecasting). The downside is that at such high altitudes, it's tough to get high-resolution images of Earth's surface. On the other hand, low Earth orbit, which can be as little as a hundred miles above Earth's surface, needs a much higher speed to keep the satellite in space. That

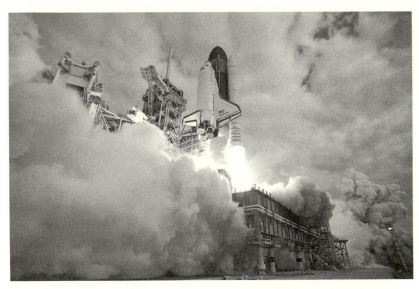

**8.4.** The space shuttle *Endeavour* carried the Shuttle Radar Topography Mission (SRTM) into space in February 2000. Image: NASA.

means it can't keep a fixed position relative to the ground like a geostationary orbit does, instead passing over any given location in a matter of minutes. However, the satellite and its imaging instruments are much closer to Earth's surface, permitting far better-quality pictures. Low Earth orbits are used by manned spacecraft like the space shuttle and International Space Station, environmental monitoring satellites like the long-running series of LandSat satellites, and spy satellites.

Satellite remote sensing provides a unique bird's-eye view (or perhaps extraterrestrial's-eye view) of river basins, giving a critically valuable, very large scale perspective we couldn't experience otherwise. Satellite images of various watershed features have appeared throughout this book, such as the NASA image of the Gangotri Glacier in the Himalayan headwaters of the Ganges River, offered earlier in this chapter. However, there are also more specific and technical applications of space-based environmental imaging. Let's give a couple examples. Consider the Shuttle Radar Topography Mission (SRTM). Consisting of a 60-meter-long radar system affixed to the space shuttle, over an eleven-day period it mapped out Earth's surface elevation at an unprecedented level. This data set has been extensively used in environmental studies, from modeling the past and future evolution of glaciers, to delineating the boundaries of watersheds. Another great example is the MODIS instrument. Shorthand for moderate-resolution imaging spectroradiometer, MODIS is a passive EM sensor that monitors several distinct bands in the visible and infrared. There are two MODIS instruments, each aboard a different satellite: Aqua and Terra. Using algorithms that capitalize on the fact that different surfaces—say, trees versus water—reflect EM radiation differently in different bands, forming material-specific signatures, MODIS data can be used to track land surface characteristics. An important application is to map snow cover across a watershed, day after day. A great deal of R&D still needs to be done to improve the ease and effectiveness with which data types like these can be incorporated into quantitative hydrological studies. But already today, MODIS snow cover data are routinely ingested into numerical prediction systems used for operational reservoir inflow forecasting, for example. We opened this chapter by noting that to understand the river at your feet, sometimes you have to look up to the stars. We could equally well say that to understand the river at your feet, sometimes you have to go the stars—and then look back down.

# 9

## THE HYDROLOGIST'S FINAL EXAM: WATERSHED MODELING

The sciences do not try to explain, they hardly even try to interpret, they mainly make models. By a model is meant a mathematical construct which, with the addition of certain verbal interpretations, describes observed phenomena. The justification of such a mathematical construct is solely and precisely that it is expected to work.
—John von Neumann

We don't absolutely require mathematical models in order to do good science—for instance, tremendous progress was made in biology and geology without the benefit of calculus—but such models nevertheless tend to be diagnostic of what I'll loosely call the maturity of a field of study. By maturity I mean the duration and intensity with which the field has been studied, the amount and depth of knowledge thus generated, and how these stack up against that field's inherent level of difficulty. The ability to predict the state of a system is a demanding test of one's knowledge of that system, and the precision of a mathematical model offers up the toughest form of such examinations. Putting it another way, an adage I've heard attributed to Ronald Reagan posits that specificity is the key to credibility. The accuracy of vague, arm-waving predictions can be tough to objectively verify—hence what may appear to be either the magical accuracy or complete uselessness of horoscopes, depending on whom you ask. Quantitative predictions do not, in principle, show any such ambiguities. Either you got the right number or you didn't. In practice, of course, the more common result is that you got a number that's pretty close but not quite right, reflecting only partial knowledge of the system, and sometimes, different models can give a similar result for different reasons, a phenomenon called nonuniqueness or equifinality. But still, the performance of such a model can be objectively assessed, and importantly, the relative performances of competing models can be readily judged. A good deal of modern science can be viewed as a horse race between competing mathematical models of a given system or phenomenon.

But recall that in the foregoing definition of maturity I included the inherent difficulty of the field. Some fields have proven remarkably resistant to broadly successful mathematical model building, in spite of the immense talents and resources devoted to them; much progress has been made, but knowledge remains incomplete. Weather forecasting is a great example—it's come a long way, it works, and it has saved countless lives and untold dollars, but such forecasts still only go a few days out, and even then they certainly aren't entirely reliable. Yet in response to such challenges, not only have efforts been redoubled, but brand new forms of mathematical analysis and modeling have also been created, representing entirely new forms of scientific understanding. Examples include chaos theory, information theory, and fractal mathematics, each discussed in other chapters of this book. These new methods and types of understanding often focus not quite so much on predictions of the exact state of a given system at a given time at a given place. Rather, they consider the overall patterns created in and by the system, generating precise mathematical descriptions and explanations of the overall look and feel of a thing: the branching, tree-like pattern of rivers and tributaries as seen on a topographic map or aerial photo (fractal geometry), or how weather systems repeat themselves over and over again but never exactly (chaos theory), or the commonness of small streamflow variations and comparative rarity of catastrophic floods (power-law scaling).

So that's the broad context. But how does it apply to watershed modeling? How well do watershed models integrate and synthesize all the hydrological considerations we've talked about in this book? How effectively do they test what we think we know about rivers? If the ability to accurately predict system behavior using quantitative models is the supreme test of scientific knowledge—the final exam, as it were, for scientists in a given field—what grades are hydrologists getting? How mature a field is hydrology? What are these models actually used for in the real world, and just what exactly is a watershed model, anyway?

It turns out that the answers to these questions are a little murkier and more open-ended that you might expect. But by briefly exploring the topic of watershed models in this penultimate chapter, we can start to gauge what the science of hydrology has accomplished, and what challenges remain unresolved—an important point, as these models form the basis for much of our predictive ability and therefore our capacity to anticipate, prevent, or manage the impacts of environmental variability and change. These models can be used for everything from hourly flood forecasting in support of emergency management, to century-long projections of climate change impacts in support of adaptation strategy development, seasonal water supply prediction to facilitate optimal management of hydroelectric and water supply reservoirs, assessing the aquatic habitat implications of a proposed development project like a mine or a subdivision, creating best management practices for agriculture, and building an improved fundamental understanding of how watersheds work. That is, watershed models are not only a hydrologic scientist's ultimate test, as explained above, but also her or his ultimate tool. So, let's jump in and take a closer look.

In principle, a watershed model is a direct or indirect mathematical and, ultimately, software representation of all the processes and characteristics of an upstream drainage area that combine to control the water resources in that watershed. This definition is extremely broad, however, and in practice, most watershed models have a somewhat narrower mission and approach. Even so, there is great variation among models. Some models include certain watershed-scale atmospheric processes, like predictions of spatial variability in the amount of precipitation and whether it falls as rain or snow; others do not. Some models include full mathematical representations of groundwater dynamics, whereas others simply "parameterize" these processes by collapsing the detailed physics into simplified approximations and parameters. Some models include sophisticated river hydraulics components, capturing detailed patterns of water depth and flow dynamics within the stream channel; many do not. Some include water quality modeling capabilities, and most don't. You get the picture. The one thing they all have in common—and this is the most fundamental function of a watershed model—is to predict streamflow on the basis of data summarizing the watershed's internal properties and the meteorological inputs to it. There are many ways to accomplish even this goal, though, and below we'll provide an abbreviated taxonomy of the ecosystem of watershed models, with some explanation of how they work and what we can and can't realistically expect them to do for us.

There are two general ways of bringing together all the ideas we have about how watersheds function into a single, unified mathematical representation and software application, with the goal of successfully reproducing and predicting river flows. One is to use direct representations of the underlying physics. Another is to use more implicit modeling methods. Let's start with the latter. These methods, which we'll call empirical, are data-driven. In an empirical model, we make little attempt to explicitly capture the physical elements of the complex causal chain that, in the end, controls river flows. Rather, we just try to find mathematical relationships that are effective at "mapping" the output—typically, streamflow—back to various potential inputs like rainfall, snow depth, temperature, the previous day's streamflow, and so forth. Classical statistical approaches, which go by names like regression, generalized linear models, principal components regression, or Box-Jenkins models, are one common approach. Many, if not most, of the operational systems for seasonal water supply forecasting in western North America use such methods, for example. Another type of empirical model is the soft-computing approach. These methods use biological systems or modes of human thought as the basis, or at least the inspiration, for universal quantitative modeling techniques. By far the most common example in hydrology is so-called machine learning, a branch of artificial intelligence, particularly in the form of the neural networks we discussed earlier in this book. Another form of soft computing is genetic programming, which involves treating the model as an organism that evolves in such a way as to do the best job possible of modeling the data. It uses automated development of an evolutionarily "fittest" mathematical model through computational representations of such genetics-inspired operations as reproduction and mutation, where the thing being evolved is the form and parameters of the mathematical model, and genetic fitness is defined as the ability to best reproduce the observed data. And then there's fuzzy logic, the aim of which is to represent human ideas of language and classification mathematically. For instance, a fuzzy inference system might classify predictors like temperature and rainfall into intuitive sets—such as low, medium, high, and very high—with overlap between adjacent sets, reflecting the fact that one person might call a certain rainfall value "medium" whereas the next person might call it "high." Then a set of rules, reflecting a codification of expert human knowledge, is used to combine classifications of predictors into set memberships for streamflow—which can similarly fall under more than one classification simultaneously. Finally, this fuzzy set membership is "defuzzified" using an algorithm into a so-called crisp number: the precise, quantitative, conventional streamflow prediction. All these soft computing techniques can also be combined with each other. Neurofuzzy models, for instance, combine artificial neural networks with fuzzy expert systems, and evolutionary neural networks integrate ideas from genetic programming, such that the neural network picks its own, optimal form. Still more concepts can be pulled into the mix, like information theory (also discussed in an earlier chapter), which is sometimes used as a basis for picking the best model form, or for choosing the best model from several competing candidates.

These empirical, data-driven models are both powerful and flexible. Granted, their successful implementation requires some physical understanding of watershed dynamics, and when the results are carefully analyzed they can also provide new understanding of those dynamics. Still, nothing about the tangible physical reality of the watershed is explicitly incorporated into the model. These techniques might just as well be applied to modeling house prices, say, or the relationship between donut intake and waist size. This is, in many ways, a good thing. Statistical and soft-computing methods amount to a form of data mining, searching out whatever patterns best associate system inputs with system outputs, and thus are equally well applied to macroeconomics or population biology or astrophysics. They let the data do the talking. They are comparatively unconstrained by incomplete physical knowledge of the system to be modeled, as such knowledge is not directly integrated into the model anyway. Similarly, they are conducive to objective model building, as the understanding or preconceptions that the scientist or engineer may have about the system play a relatively minor role. And they are also, very often, the most accurate, efficient, and cheapest predictive tools available. That makes them particularly well suited to tasks where performance is paramount, like short-term flood forecasting, which often forms a direct input to the decision whether to evacuate an area, for example.

But to every yin, there's a yang. In some sense, these methods bypass one of the main uses of mathematical-computational models: to test one's knowledge of the physical processes that go together to create the system dynamics. If no such knowledge is explicitly incorporated into the model, the model cannot be used to explicitly test that knowledge. And there's a more practical downside, too. What if, for example, we want to explore how flows in a given stream might be affected by a proposed housing development within the watershed? Because this sort of model is more or less a black box, you generally can't explore what implications new environmental circumstances might incur by just fidgeting with its internal parts. This disconnect between the inner workings of (for instance) a machine learning system versus our intuitive human understanding of the thing being modeled is by no means unique to hydrologic contexts. Consider the following example, drawn from an article by Steven Levy in *Wired*, which talks about the ubiquity of both artificial intelligence and robotics technologies in our everyday lives—in this case, a baby products warehouse—and how that has turned out rather differently from how science fiction movies of the past often imagined it:

> Boxes of pacifiers sit above crates of onesies, which rest next to cartons of baby food. In a seeming abdication of logic, similar

items are placed across the room from one another. A person trying to figure out how the products were shelved could well conclude that no form of intelligence—except maybe a random number generator—had a hand in determining what went where . . . but the warehouses aren't meant to be understood by humans; they were built for bots. Every day, hundreds of robots course nimbly through the aisles, instantly identifying items and delivering them to flesh-and-blood packers on the periphery . . . [these] bots may not seem very smart. They don't possess anything like human intelligence and certainly couldn't pass a Turing test. But they represent a new forefront in the field of artificial intelligence. Today's AI doesn't try to re-create the brain. Instead, it uses machine learning, massive data sets, sophisticated sensors, and clever algorithms to master discrete tasks.

Trying to interpret, in terms of specific physical hydrologic processes, the internal architecture of a neural network optimized for predicting streamflow would be kind of like trying to interpret the AI-controlled operations of that warehouse. More challenging still would be an attempt to identify and manipulate some aspect of the detailed inner workings of the neural network so as to simulate the potential effect of a proposed new pumping well upon groundwater inflows to an adjacent salmon stream, say.

For this, we would instead need the second major type of watershed model. Process-oriented models directly and explicitly represent the salient hydrological processes and characteristics of a watershed. More specifically, process-oriented models are sets of mathematical equations, normally implemented as computer code, that seek to clearly and specifically represent the relevant physical processes occurring in a watershed. They usually consist of various components or modules—for instance, equations representing evaporation, equations representing infiltration of rainwater into soil, equations representing uptake of soil moisture by forests or crops, equations representing the formation of snowpack in winter and its melting in spring. Process-oriented models aren't generic black boxes, like the empirical models discussed above. Rather, they're purpose-built as virtual-reality watersheds—rivers in silicon, as it were. And because the various components of the terrestrial hydrologic cycle are explicitly specified, we can change their properties around and see what happens, conducting numerical experiments upon our virtual watershed. For example, the area of the watershed covered by forest canopy might be directly fed to the model—so we could change it to reflect deforestation, say, and then rerun the model to estimate how that might change the dynamics of streamflow generation. And by the same token, because our current understanding of watershed dynamics is specifically and explicitly encoded in the model, by seeing how well the model reproduces reality we can test how accurate that understanding really is.

That said, there's no one right way to build a process-oriented watershed model either. There is no "standard model" of watershed hydrology akin to that in particle physics, for instance. A review paper in the *Journal of Hydrologic Engineering* once cataloged over sixty different process-oriented models. Broadly speaking, these differ in terms of the size of the watersheds they're optimized for simulating, the level of spatial detail they can incorporate, the specific physical processes they capture, the amount and type of input data they require, and their overall ease, speed, and cost of use. Let's list a few examples just to get a feeling for this. The Distributed Hydrology Soil Vegetation Model (DHSVM), developed at the University of Washington, performs highly detailed modeling of forest hydrology processes in small watersheds. The University of British Columbia Watershed Model (UBCWM), developed at its namesake institution, and the Hydrologiska Byråns Vattenbalansavdelning (HBV) model, developed at the Swedish Meteorological and Hydrological Institute, were designed for streamflow simulation in data-sparse mountain environments, and they are among the handful of watershed models that explicitly incorporate glacier melt as a separate physical phenomenon from seasonal snowmelt. The Système Hydrologique Européen (SHE), developed by a joint British-French-Danish research consortium, is one of very few streamflow models to directly solve the differential equations governing watershed hydrology using numerical methods of the sort we briefly mentioned in our discussion of calculus. The Soil and Water Assessment Tool (SWAT) was originally developed by the US Department of Agriculture for investigating the impacts of land management practices and considers both streamflow and water quality. The list goes on—and as you can see, any discussion of that list involves a litany of acronyms!

So how does one choose among all these approaches to hydrologic modeling? As in so many fields of human endeavor, from rebuilding a car engine to addressing a breach of international law, it all comes down to picking the right tool for the job. We already discussed some model selection criteria above. Empirical models tend to provide great bang for the buck: they're very flexible, relatively low cost to build and run, and often provide superior simulation performance. Such predictive accuracy is really important if you happen to be responsible, for example, for telling your local emergency management agency whether a flood is imminent and the townsfolk need to clear out. On the other hand, process-oriented models can be a better choice if we want to investigate the impacts of environmental change, objectively test our scientific knowledge of watershed processes, or perform diagnostics—a common example of the latter being the retrospective studies of detailed cause and effect that often follow on the heels of major storm systems, especially if the resulting flood damage was worse than expected. Furthermore, selecting a particular modeling technology from one of these two broad classes will typically involve application-specific considerations. For example, a conventional linear regression model can be built by anyone with a good working knowledge of basic statistics, using spreadsheet software ubiquitous in the modern office, whereas a genetic programming model requires significantly greater institutional capacity. And if you'd like to conduct research on the effects of different forestry practices on aquatic habitat availability, DHSVM could be a good option, but you might prefer the HBV model if you instead wanted to look at how downstream water resource availability could be impacted by glacier recession under climate change.

So if mathematical models are the ultimate test of scientific knowledge, just how well are hydrologists doing on their final exam? A bit of a mixed result, it seems. Some questions in the hydrologist's final exam are easier than others. We can usually do a good job of reproducing, and in many cases even making seasonal forecasts for, total yearly streamflow volumes. This is a matter of closing the large-scale water budget—balancing the total inputs, like precipitation, against the total outputs, like evaporation and streamflow—and this exercise is comparatively robust to incomplete scientific knowledge and mediocre data. When we look at smaller spatial scales, or in particular, at temporally finer-scale dynamics—peak flood flows, for example, or very low flows—these remain more challenging to reliably reproduce, because we need lots of data and a very good mathematical representation of what's going on in the watershed.

Let's examine this point a little more closely. As one example, consider the rain-on-snow events that often lead to extreme floods in mountain regions. These events are a "perfect storm" in which two main factors come together: snow on the ground, and a warm, moist weather system, such as the Pineapple Express storms we discussed earlier in this book. Runoff is then generated by simultaneous rapid snowmelt and intense rainfall.

However, the effects can be highly localized. For instance, snow may only reside in the upper reaches of the watershed, so that rain-on-snow processes occur over only part of the model domain. The algorithms that watershed models use for seasonal snow accumulation and melt are imperfect, and the detailed data required to constrain the exact locations where snow is or isn't, and the state of the snow—its temperature, and saturation by liquid water, are important for determining how strongly the snowpack will respond to a rain-on-snow event— are typically unavailable. And there are still more complications, such as knowing the exact distribution of rainfall intensity in a mountainous environment, where rainfall can be very variable, and little data exist. Desert environments are another example of a hydrological landscape that poses significant challenges for hydrologists. Ephemeral stream channels, also known as dry washes, arroyos (Spanish), and wadis (Arabic), are the predominant surface water feature of deserts. They are susceptible to flash floods, going from bone dry to a raging torrent, sometime in mere minutes. This reflects the great intensity of rainfall in the desert when it does occur, and in some cases the impermeability of desert soils due to formation of a hard calcium carbonate crust called caliche, which limits infiltration. Think back to our chapter on streamflow memory: because water infiltration and storage can be limited, desert streams are often low-memory systems, tightly coupled to the unpredictable weather overhead, a little like an urban landscape. Moreover, a flood may happen at a downstream location where it isn't even raining, due to an intense thunderstorm in the watershed's mountain headwaters, which can make it easy to be caught unaware—part of the reason such flash floods can be so deadly. Such erratic streams can be hard to model well.

A major path forward in hydrological modeling today is the use of Monte Carlo techniques. The term was coined by the mathematician Stanisław Ulam, who evidently came up with

**9.1.** Certain topics present particularly tricky questions for watershed modelers. Above: a dry wash near Las Vegas. Photo: S. W. Fleming.

the method while working on the Manhattan Project. He used it to solve differential equations, a little like the advection-dispersion equation we introduced in our discussion about groundwater contaminant transport. But the idea as it's widely applied to simulation modeling throughout the sciences and engineering—an application that has diffused into hydrological modeling as well—is to estimate model uncertainty by generating not one, but many, streamflow predictions using randomly (but systematically) perturbed model inputs or parameters. That is, the watershed model is simply run over and over and over again in slightly different ways, each of which is reasonable, producing slightly different output each time. The technique can be expanded to include a multiplicity of different watershed models, and in streamflow forecasting applications, it can even include the predictions from many different weather models as the inputs to many hydrologic models. The output from each of these model runs is independently stored, and the end result is a collection, or "ensemble," of output hydrographs—potentially, thousands of them. The ensemble as a whole gives an envelope of possibilities. If done well, that envelope should capture reality somewhere within it, and further, the probabilities of different outcomes can be calculated. For example, using such methods, you might predict that there's a 92% chance that the river will top its banks within the next twenty-four hours. While it may seem like gambling—hence the name—Monte Carlo simulation is a very useful way of addressing uncertainty in environmental predictions. In fact, much of the focus over the last decade or two has been not so much on attempting to improve the models per se, but instead on improving estimates of uncertainty in the best available prediction. From the perspective of someone who uses hydrologic simulations as input to real-world decision processes, such as an environmental manager or design engineer, it's really good to know what the "error bars" around the predictions are. An interesting addendum is that, in general, the best watershed model of all is a bunch of watershed models. That is, if we take all these outputs and average them together, the resulting ensemble mean typically outperforms even the best individual ensemble member. The reason for this counterintuitive result is that, as any model is a simplified representation of reality, all models are at least partly wrong—but because different models typically have different errors, the mistakes can mutually cancel each other out when you average together the results from (a sufficiently large number of) different models.

And what of the new horizons in quantitative analysis summarized at the start of this chapter, the fractal view of nature for instance, or the chaos interpretation of nonlinear dynamical systems like the weather? What role do these play in applied watershed modeling? Here we again have something of a mixed bag. Certainly, few watershed models directly incorporate any of these concepts at all, and that is unfortunate if not altogether dismaying. In other ways, however, these ideas permeate the field. Any kind of model-based river forecasting is hugely impacted by uncertainty in future weather, which in turn stems from deterministic chaos, for instance, and every hydrologist should be aware of the fractal spatial pattern of most stream networks.

In fact, one of the main challenges for future hydrologists— their real final exam—may be to find new ways to take sophisticated descriptive measures of complexity and apply them in a workmanlike and useful manner within watershed models, to better capture the scientific realities of the Earth system and to form ever more accurate river predictions for the benefit of all watershed stakeholders.

# 10

## EPILOGUE

The importance of bringing water resource science to the next level will only grow. Today, according to a United Nations report, more than one billion people already do not have access to clean water, and 1.4 billion live in river basins where water use exceeds recharge rates. Another two billion or so water users will be added to the world's population by midcentury. This population growth, together with expansion of agricultural and industrial production as poorer nations develop, are expected to increase the global water demand by a stunning 55% by 2050. There is, in short, more and more pressure on the resource.

Not only do these factors increase water demand, they also signify greater global exposure to water-related hazards, including pollution and flood risk, as more people settle on floodplains, for instance, and more industrial and agricultural effluent is discharged into the environment. At the same time, the current scientific consensus is that the net increase in atmospheric greenhouse gas concentrations resulting from activities like massive fossil fuel combustion, industrial livestock production, and widespread deforestation is sufficiently large to detectably alter global climate. Current projections suggest that the main hydrological effect for most basins will be to amplify the water cycle, which may actually result in an increase in runoff and potentially available water supplies in many regions, but which may reduce supplies in others. More importantly, it may widely increase the intensity of both the yearly rainy season and yearly dry season, further increasing flood and drought risks. And river channelization, damming, contamination, and upstream water withdrawals have so degraded aquatic habitat that many freshwater biological populations have collapsed, some species have been entirely extirpated from parts of their home ranges, and others are at risk of extinction altogether. We are facing a dark constellation of regional water resource disasters, growing and coalescing into what appears to be an emerging global catastrophe of human welfare and the environment.

I am nevertheless cautiously optimistic. Granted, the fundamental driving forces behind our water resource challenges are not really on the table for discussion. Virtually every society pursues continuous growth of its population and its economy. Rate and type of growth may be debated, but growth it will be. And though some irresponsible growth policies can be stopped, overall, there's only so much that we can do about this within a context of democratic governance. The first of only two convincing threads of genuine systematic progress in human civilization over the centuries (we'll get to the second very shortly) is the emergence of individual rights and freedoms. No movement, no matter how lofty in its intentions or confident of its facts, has the moral authority to control a person's choice to raise a family or to earn wealth by the honest sweat of his or her own brow. So, why the optimism, then?

I posit that the answer lies with the second of those two aforementioned threads of consistent progress in civilization: science. Assuming that strong demographic and economic self-restraint are nonstarters, we evidently have three options. These choices are, first, Malthusian population collapse (generally felt to be unlikely, but you never know). The second is a sort of dynamical equilibrium between ever-increasing human populations and our incrementally improving ability to just barely keep them fed and watered, forming a race to the bottom, which, in effect, maximizes the net global misery (more or less the status quo). The third is to invest deeply in a coordinated, broad-based, and large-scale drive to create new science and technology that addresses the needs and aspirations of the current and future global populations in a healthy and sustainable way. Personally, I vote for option number three. It will be challenging, there will be unintended consequences, and we must reject any attempt to equate technology with technocracy: at an individual level, everyone must have the right, capacity, and opportunity to turn the technology off. But that said, the rewards will be worth the effort.

There is justification for having some faith that we might be able to get real traction on water resources issues. The development of ever more water-efficient technologies for homes, farms, and factories is an obvious example. Indeed, water use in the United States has leveled off near 1970 rates in spite of both population and economic growth. To be sure, unsustainable water practices during regional droughts, such as groundwater mining in California at the time of writing, reveal a chink in the armor. Furthermore, such stabilization of water demand currently seems restricted, at best, to a handful of rich nations. Nevertheless, the overall statistic must be acknowledged as the stunning success, cause for optimism, and clear template for emulation that it is—a shining citadel on the hill, as it were. Improvements in water quality are another example. Granted, in many parts of the world, rapid agricultural and industrial expansion are making water quality worse, not better, and nitrate contamination from the overuse of fertilizer seems to be an especially intractable problem. Shortages of potable water due to fecal contamination remains a huge issue globally, and emerging contaminants, like pharmaceuticals, are an increasingly worrisome threat to ecosystems and water and food supplies. Yet improved awareness, legislation, and technology have resulted in tremendous gains. The days of rivers ablaze—this happened to the Cuyahoga River in Ohio, which was so polluted with flammable contaminants that in 1969 it actually caught fire—seem to be over. Overall, water quality across the industrialized regions of the developed world may actually be much better now than it was, say, forty or fifty years ago. Another broad reason for optimism is the seeming paradox of water conflict. Many have predicted "water wars," and Ismail Serageldin, a former World Bank vice president, warned that the wars of the twenty-first century will be fought over water. It turns out, though, that water resource conflict and cooperation are surprisingly nuanced. Wars solely over water, even in regions that are both arid and troubled, are virtually nonexistent. There are even examples of nations, at war over other issues, which have met peacefully to allocate transboundary water resources. Consider the River Jordan, pictured in figure 10.1, which is shared by Syria, Jordan, and Israel. When you hear on the news of the "West Bank" part of the Palestinian territories, you're actually hearing a reference to the western bank of the River Jordan. It forms the border between Israel and Jordan, the Sea of Galilee lies in its headwaters, and it flows to the Dead Sea. It is, in short, at the geographic center of a great deal of history, not to mention strife. In 1965, when Syria was building an upstream diversion of a tributary to the River Jordan that would reduce Israel's water supply by 11%—a catastrophe for a desert nation—Israel responded with air strikes against the facility. These events were a contributing factor to the Six-Day War a couple years later and might seem to be convincing evidence that water wars are a stark reality. However, competition over scarce water was only one of several factors leading to war. And more importantly (and interestingly), Israel and Jordan held ongoing secret "picnic table" talks to peacefully manage the Jordan starting in 1953—even though they were officially at war from 1948 until the Israel-Jordan Peace Treaty of 1994.

10.1. A tank being moved across the River Jordan by Israeli Defense Forces. Photo by Cpl. Shay Wagner, IDF Spokesperson's Unit.

Globally, the pattern seems to be that disagreements over water are ubiquitous; violent civil conflict over water occurs with some regularity but is uncommon overall; and violent international conflict over water is exceedingly rare. At all levels, cooperation might be at least as common as conflict—though as global demand explodes, this all might change.

As water pressures mount, the need to predict and understand the hydrological cycle grows commensurately. Water isn't optional. Water is necessary for our very existence, for our continued economic development, and for the health of the web of life that supports us. It's also limited in its availability, and there are no substitutes for it. Whatever path humanity chooses to follow, it will be up to hydrologists to present society with the options available, and the corresponding pros and cons, for the management of our water resources. And to do that, hydrologists need to understand watershed systems in detail, and to accurately, precisely, consistently, and quantitatively predict the impacts on those systems from both natural and human interventions. Remember again the quote on models from von Neumann, given at the start of the preceding chapter: *the justification of such a mathematical construct is solely and precisely that it is expected to work.* We can only achieve the required level of performance by assembling new generations of mathematical-computational models, exhibiting far greater sophistication and imagination, along with—it must be emphasized—the observational data required to build and test them robustly.

Viewed from this perspective, it is perhaps not too dramatic to assert that the future of the world will depend, in a small but real way, on a quantum leap forward in our understanding of the physics of rivers.

# SOME REFERENCES
# AND SUGGESTIONS FOR
# FURTHER READING

### CHAPTER 1. INTRODUCTION

Dingman, S. L. 1994. *Physical Hydrology*. Englewood Cliffs, NJ: Prentice-Hall.

Maclean, N. 1976. *A River Runs through It, and Other Stories*. Chicago: University of Chicago Press.

Reid, M. 1991. *Myths and Legends of the Haida Indians of the Northwest*. Santa Barbara, CA: Bellerophon.

United Nations Development Programme. 2006. *Beyond Scarcity: Power, Poverty and the Global Water Crisis, Human Development Report 2006*. New York: UNDP.

### CHAPTER 2. WHY RIVERS ARE WHERE THEY ARE

Armstrong J. E. 1990. *Vancouver Geology*. Vancouver, BC: Geological Association of Canada, Cordilleran Section.

Badash, L. 1968. "Rutherford, Boltwood, and the Age of the Earth: The Origin of Radioactive Dating Techniques." *Proceedings of the American Philosophical Society* (112): 157–69.

Cannings, S., and R. Cannings. 1999. *Geology of British Columbia: A Journey through Time*. Vancouver, BC: Greystone.

Clague, J., and B. Turner. 2003. *Vancouver, City on the Edge: Living with a Dynamic Geological Landscape*. Vancouver, BC: Tricouni.

Crutzen, P. J. 2002. "Geology of Mankind." *Nature* (415): 23.

Dietz, R. S. 1961. "Continent and Ocean Basin Evolution by Spreading of the Sea Floor." *Nature* (190): 854–57.

Fleming, S. W., and A. M. Trehu. 1999. "Crustal Structure beneath the Central Oregon Convergent Margin from Potential-Field Modeling: Evidence for a Buried Basement Ridge in Local Contact with a Seaward Dipping Backstop." *Journal of Geophysical Research* (104): 20431–47.

Hess, H. H. 1962. "History of Ocean Basins." In *Petrologic Studies: A Volume in Honor of A. F. Buddington*, edited by A.E.J. Engel,

H. L. James, and B. F. Leonard, 599–620. Boulder, CO: Geological Society of America.

Leopold, L. B. 1994. *A View of the River*. Cambridge, MA: Harvard University Press.

Ludvigsen, R., and G. Beard. 1997. *West Coast Fossils: A Guide to the Ancient Life of Vancouver Island*. Madeira Park, BC: Harbour.

Mathews, W. H., and J.W.H. Monger. 2005. *Roadside Geology of Southern British Columbia*. Missoula, MT: Mountain Press.

Monger, J.W.H., ed. 1994. Geology and Geological Hazards of the Vancouver Region, Southwestern British Columbia, GSC Bulletin 481. Ottawa, ON: Geological Survey of Canada.

Orr, W. N., and E. L. Orr. 2002. *Geology of the Pacific Northwest*. 2nd ed. New York: McGraw-Hill.

Pelletier, J. D. 2008. *Quantitative Modeling of Earth Surface Processes*. Cambridge: Cambridge University Press.

Vine, F. J., and D. H. Matthews. 1963. "Magnetic Anomalies over Ocean Ridges." *Nature* (199): 947–49.

Wegener, A. 1915. *Die Entstehung der Kontinente und Ozeane*. Braunschweig: Vieweg.

Wilson, J. T. 1963. "Evidence from Islands on the Spreading of Ocean Floors." *Nature* (197): 536–38.

Yergin, D. 1991. *The Prize*. New York: Simon and Schuster.

CHAPTER 3. HOW DO RIVERS REMEMBER?

Bass, T. A. 1999. *The Predictors*. New York: Henry Holt.

Beran, J. 1994. *Statistics for Long-Memory Processes: Monographs on Statistics and Applied Probability*, vol. 61. New York: Chapman and Hall.

Chatfield, C. 1996. *The Analysis of Time Series*. 5th ed. London: Chapman and Hall.

Dai, A., T. Qian, and K. E. Trenberth. 2009. "Changes in Continental Freshwater Discharge from 1948 to 2004." *Journal of Climate* (22): 2773–92.

Dingman, S. L. 1994. *Physical Hydrology*. Englewood Cliffs, NJ: Prentice-Hall.

Fleming, S. W. 2007. "Artificial Neural Network Forecasting of Nonlinear Markov Processes." *Canadian Journal of Physics* (85): 279–94.

Fleming, S. W. 2007. "Quantifying Urbanization-Associated Changes in Terrestrial Hydrologic System Memory." *Acta Geophysica* (55): 359–68.

Gleick, J. 1987. *Chaos: Making a New Science*. New York: Penguin.

Hurst, H. E. 1957. "A Suggested Statistical Model of Some Time Series That Occur in Nature." *Nature* (180): 494.

Konrad, C. P., D. B. Booth, and S. J. Burges. 2005. "Effects of Urban Development in the Puget Lowland, Washington, on Interannual Streamflow Patterns: Consequences for Channel Form and Streambed Disturbance." *Water Resources Research* (41), doi:10.1029/2005WR004097.

Lorenz, E. N. 1963. "Deterministic Nonperiodic Flow." *Journal of the Atmospheric Sciences* (20): 130–41.

Rikitake T. 1958. "Oscillations of a System of Disk Dynamos." *Mathematical Proceedings of the Cambridge Philosophical Society* (54): 89–105.

Toussoun, O. 1925. "Chapitre XI: Les Niveaux, Tome Deuxiéme: Memoire sur l'Histoire du Nil." In *Memoires à L'Institut D'Egypte*, 366–92. Cairo: Imprimerie de l'Institut Français d'Archéologie Orientale, Cairo, Egypt.

CHAPTER 4. CLOUDS TALKING TO FISH:
THE INFORMATION CONTENT OF RAIN

Adams, D. 1979. *The Hitchhiker's Guide to the Galaxy*. London: Pan Books.

Bell, A.J.W., director and producer. 1981. *The Hitchhiker's Guide to the Galaxy*. United Kingdom: BBC 2.

Caselton, W. F., and T. Husain. 1980. "Hydrologic Networks: Information Transmission." *ASCE Journal of the Water Resources Planning and Management Division* (106): 503–20.

Dodds, W. K. 2002. *Freshwater Ecology*. San Diego, CA: Academic Press.

Fleming, S. W., and D. J. Sauchyn. 2013. "Availability, Volatility, Stability, and Teleconnectivity Changes in Prairie Water Supply from Canadian Rocky Mountain Sources over the Last Millennium." *Water Resources Research* (49): 64–74.

Fleming, S. W., and F. A. Weber. 2012. "Detection of Long-Term Change in Hydroelectric Reservoir Inflows: Bridging Theory and Practise." *Journal of Hydrology* (470–71): 36–54.

Gleick, J. 2011. *The Information: A History, a Theory, a Flood*. New York: Vintage.

Konrad, C. P., D. B. Booth, and S. J. Burges. 2005. "Effects of Urban Development in the Puget Lowland, Washington, on Interannual Streamflow Patterns: Consequences for Channel Form and Streambed Disturbance." *Water Resources Research* (41), doi:10.1029/2005WR004097.

Krasovskaia, I. 1995. "Quantification of the Stability of River Flow Regimes." *Hydrological Sciences Journal* (40): 587–98.

Pierce, J. R. 1980. *An Introduction to Information Theory: Symbols, Signals, and Noise*. 2nd ed. New York: Dover.

Rosales J., L. Blanco-Belmonte, and C. Bradley. 2007. "Hydrogeomor-phological and Ecological Interactions in Tropical Floodplains: The Significance of Confluence Zones in the Orinoco Basin, Venezuela." In *Hydroecology and Ecohydrology: Past, Present, and Future*, edited by P. J. Wood, D. M. Hannah, and J. P. Sadler, 295–316. Chichester, UK: John Wiley and Sons.

Shannon, C. E. 1948. "A Mathematical Theory of Communication." *Bell System Technical Journal* 27 (3): 379–423.

———. 1948. "A Mathematical Theory of Communication." *Bell System Technical Journal* 27 (4): 623–66.

Shannon, C. E., and W. Weaver. 1949. *The Mathematical Theory of Communication*. Urbana: University of Illinois Press. Reprinted in 1963 and 1998.

Siegfried, T. 2000. *The Bit and the Pendulum: From Quantum Computing to M Theory—The New Physics of Information*. New York: John Wiley.

Stockton, C. W., and G. C. Jacoby Jr. 1976. "Long-Term Surface-Water Supply and Streamflow Trends in the Upper Colorado River Basin Based on Tree-Ring Analyses." Lake Powell Research Bulletin Vol. 18. University of California, Los Angeles.

## CHAPTER 5. SEARCHING FOR BURIED TREASURE

Alley, W. M., R. W. Healy, J. W. LaBaugh, and T. E. Reilly. 2002. "Flow and Storage in Groundwater Systems." *Science* (296): 1985–90.

Anderson, M. P. 2007. "Introducing Groundwater Physics." *Physics Today* (60): 42–47.

Armstrong, A. 2012. "Supply and Demand." *Nature Geoscience* (5): 592.

Barnhardt, W. A., R. E. Kayen, J. D. Dragovich, S. P. Palmer, P. T. Pringle, B. L. Sherrod, and S. V. Dadisman. 2003. "The Effects of Volcanic Debris Flows (Lahars), Earthquakes and Landslides on Holocene Deltas at Puget Sound, Washington." US Geological Survey, walrus. wr.usgs.gov/geotech/pugetposter/index.html (accessed March 15, 2016).

Domenico, P. A., and F. W. Schwartz. 1990. *Physical and Chemical Hydrogeology*. 2nd ed. New York: John Wiley and Sons.

Douglas, T. 2006. "Review of Groundwater-Salmon Interactions in British Columbia." Report prepared for Watershed Watch and the Walter and Duncan Gordon Foundation.

Fleming, S. W., and R. Haggerty. 2001. "Modeling Solute Diffusion in the Presence of Pore-Scale Heterogeneity: Method Development and an Application to the Culebra Dolomite Member of the Rustler Formation, New Mexico, USA." *Journal of Contaminant Hydrology* (48): 253–76.

Fleming, S. W., C. Wong, and G. Graham. 2014. "The Unbearable Fuzziness of Being Sustainable: An Integrated, Fuzzy Logic–Based Aquifer Health Index." *Hydrological Sciences Journal* (59): 1154–66.

Francis, D. 1997. *Bre-X: The Inside Story*. Toronto, ON: Key Porter.

Freeze, R. A., and R. L. Harlan. 1969. "Blueprint for a Physically-Based, Digitally-Simulated Hydrologic Response Model." *Journal of Hydrology* (9): 237–58.

Paterson, T. W. 2007. *Riches to Ruin: The Boom to Bust Saga of Vancouver Island's Greatest Copper Mine*. Duncan, BC: Firgrove Publishing.

Telford, W. M., L. P. Geldart, and R. E. Sheriff. 1990. *Applied Geophysics*. 2nd ed. Cambridge: Cambridge University Press.


CHAPTER 6. THE DIGITAL RAINBOW

Bartlett, A. A. 2004. "Thoughts on Long-Term Energy Supplies: Scientists and the Silent Lie." *Physics Today* (57): 53–55.

Bates, J.H.T. 1998. "Linear Systems and the 1-D Fourier Transform." In *The Fourier Transform in Biomedical Engineering*, edited by T. M. Peters and J. Williams, 25–52. Boston: Birkhäuser.

Brillinger, D. R. 1993. "The Digital Rainbow: Some History and Applications of Numerical Spectrum Analysis." *Canadian Journal of Statistics* (21): 1–19.

Cohen, J. E. 1995. "Population Growth and Earth's Human Carrying Capacity." *Science* (269): 341–46.

Colman, R., W. Collins, J. Haywood, M. Manning, and P. Mote P. 2007. "The Physical Science behind Climate Change." *Scientific American*, August, 64–71.

Fleming, S. W., and H. E. Dahlke. 2014. "Parabolic Northern-Hemisphere River Flow Teleconnections to El Niño–Southern Oscillation and the Arctic Oscillation." *Environmental Research Letters* (9): 104007.

Fleming, S. W., A. M. Lavenue, A. H. Aly, and A. Adams. 2002. "Practical Applications of Spectral Analysis to Hydrologic Time Series." *Hydrological Processes* (16): 565–74. See also erratum: 2003, *Hydrological Processes* (17): 883.

Hammond, J. C. 2014. "Trends in Streamflow above and below Dams across the Columbia River Basin from 1950 to 2012: Assessing Sub-Basin Sensitivity." Master's thesis, Oregon State University, Corvallis.

Murtaugh, P. A., and M. G. Schlax. 2009. "Reproduction and the Carbon Legacies of Individuals." *Global Environmental Change* (19): 14–20.

Peters, T. M. 1998. "Introduction to the Fourier Transform." In *The Fourier Transform in Biomedical Engineering*, edited by T. M. Peters and J. Williams, 1–24. Boston: Birkhäuser.

Peters, T. M., and J.H.T. Bates. 1998. "The Discrete Fourier Transform and the Fast Fourier Transform." In *The Fourier Transform in Biomedical Engineering*, edited by T. M. Peters and J. Williams, 175–94. Boston: Birkhäuser.

United Nations Development Programme. 2006. *Beyond Scarcity: Power, Poverty, and the Global Water Crisis, Human Development Report 2006*. New York: UNDP.

Vörösmarty, C. J., P. Green, J. Salisbury, and R. B. Lammers. 2000. "Global Water Resources: Vulnerability from Climate Change and Population Growth." *Science* (289): 284–88.

### CHAPTER 7. LANDSLIDES, FRACTALS, AND ARTIFICIAL LIFE

Bak, P., C. Tang, and K. Wiesenfeld. 1988. "Self-Organized Criticality." *Physical Review A* (38): 364–74.

Brown, J. R. 2004. *Minds, Machines, and the Multiverse*. New York: Touchstone.

Cipra, B. A. 2003. "A Healthy Heart Is a Fractal Heart." *SIAM News* 36 (7).

Cohn, T. A., and H. F. Lins. 2005. "Nature's Style: Naturally Trendy." *Geophysical Research Letters* (32): doi:10.1029/2005GL024476.

Fleming, S. W. 2008. "Approximate Record-Length Constraints for Experimental Identification of Dynamical Fractals." *Annalen der Physik* (Berlin) (17): 955–69.

Fraedrich, K., U. Luksch, and R. Blender. 2004. "1/f Model for Long-Time Memory of the Ocean Surface Temperature." *Physical Review E* (70): 037301.

Guthrie, R. H., P. J. Deadman, A. R. Cabrera, and S. G. Evans. 2008. "Exploring the Magnitude-Frequency Distribution: A Cellular Automata Model for Landslides." *Landslides* (5): 151–59.

Hurst, H. E. 1957. "A Suggested Statistical Model of Some Time Series That Occur in Nature." *Nature* (180): 494.

Huybers, P., and W. Curry. 2006. "Links between Annual, Milankovitch, and Continuum Temperature Variability." *Nature*, doi:10.1038/nature04745.

Jakob, M., J. J. Clague, and M. Church. 2015. "Rare and Dangerous: Recognizing Extra-Ordinary Events in Stream Channels." *Canadian Water Resources Journal / Revue canadienne des ressources hydriques*, doi:10.1080/07011784.2015.1028451.

Livina, V. N., Y. Ashkenazy, P. Braun, R. Monetti, A. Bunde, and S. Havlin. 2003. "Nonlinear Volatility of River Flux Fluctuations." *Physical Review E* (67): 042101.

Mandelbrot, B. B., and R. L. Hudson. 2004. *The (Mis)Behavior of Markets*. New York: Basic Books.

Mudelsee, M. 2007. "Long Memory of Rivers from Spatial Aggrega-tion." *Water Resources Research* (43), doi:10.1029/2006WR005721.

Nash, J. 1950. *Non-Cooperative Games*. PhD thesis, Princeton University.

Pelletier, J. D. 1997. "Fractal Models in Geology." PhD thesis, Cornell University, Ithaca, NY.

Press, W. H. 1978. "Flicker Noises in Astronomy and Elsewhere." *Comments on Astrophysics* (7): 103–19.

Slaymaker, O. "The Distinctive Attributes of Debris Torrents." *Hydro-logical Sciences Journal* (33): 567–73.

Turcotte, D. L. 1999. "Self-Organized Criticality." *Reports on Progress in Physics* (62): 1377–429.

Von Neumann, J., and A. W. Burks. 1966. *Theory of Self-Reproducing Automata*. Urbana: University of Illinois Press.

Wieczorek, G. F., M. C. Larsen, L. S. Eaton, B. A. Morgan, and J. L. Blair. 2001. "Debris-Flow and Flooding Hazards Associated with the December 1999 Storm in Coastal Venezuela and Strategies for Mitigation," USGS Open File Report 01-0144. Reston, VA: US Geological Survey.

CHAPTER 8. THE SKY'S NOT THE LIMIT

Dahlke, H. E., S. W. Lyon, J. R. Stedinger, G. Rosqvist, and P. Jansson. 2012. "Contrasting Trends in Hydrologic Extremes for Two Sub-Arctic Catchments in Northern Sweden: Does Glacier Melt Matter? *Hydrology and Earth System Science* (16): 2123–41.

Fleming, S. W., and G.K.C. Clarke. 2003. "Glacial Control of Water Resource and Related Environmental Responses to Climatic Warm-ing: Empirical Analysis Using Historical Streamflow Data from Northwestern Canada." *Canadian Water Resources Journal/Revue canadienne des ressources hydriques* (28): 69–86.

Fleming, S. W., G.K.C. Clarke, and R. D. Moore. 2006. "Glacier-Mediated Streamflow Teleconnections to the Arctic Oscillation." *International Journal of Climatology* (26) 619–36.

Hays, J. D., J. Imbrie, and N. J. Shackleton. 1976. "Variations in the Earth's Orbit: Pacemaker of the Ice Ages." *Science* (194): 1121–32.

Li, Z., W. Wang, M. Zhang, F. Wang, and H. Li. 2010. "Observed Changes in Streamflow at the Headwaters of the Urumqi River, Eastern Tianshan, Central Asia." *Hydrological Processes* (24): 217–24.

Marcott, S. A., J. D. Shakun, P. U. Clark, and A. C. Mix. 2013. "A Reconstruction of Regional and Global Temperature for the Past 11,300 Years." *Science* (339): 1138–201.

O'Neel S, E. Hood, A. L. Bidlack, S. W. Fleming, M. L. Arimitsu, A. Arendt, E. Burgess, C. J. Sergeant, A. H. Beaudreau, K. Timm, G. D. Hayward, J. H. Reynolds, and S. Pyare S. 2015. "Icefield-to-

Ocean Linkages across the Northern Pacific Coastal Temperate Rainforest Ecosystem." *Bioscience* (65): 499–512.

Tol, R.S.J., and A. Langen. 2000. "A Concise History of Dutch River Floods." *Climatic Change* (46): 357–69.

Trubilowicz, J. W., E. Chorlton, S. J. Déry, and S. W. Fleming. 2015. "Satellite Remote Sensing for Water Resource Applications in British Columbia." *Innovation, Magazine of the Association of Professional Engineers and Geoscientists of British Columbia* (March/April): 18–20.

Van Dijk, A.I.J.M., L. J. Renzullo, Y. Wada, and P. Tregoning. 2014. "A Global Water Cycle Reanalysis (2003–2012) Merging Satellite Gravimetry and Altimetry Observations with a Hydrological Multi-Model Ensemble." *Hydrology and Earth System Sciences* (18): 2955–73.

CHAPTER 9. THE HYDROLOGIST'S FINAL EXAM:
WATERSHED MODELING

Beven, K. 2001. *Rainfall-Runoff Modelling: The Primer*. Chichester, UK: Wiley.

Bourdin, D. R., S. W. Fleming, and R. B. Stull. 2012. "Streamflow Modelling: A Primer on Applications, Approaches, and Challenges." *Atmosphere-Ocean* (50): 507–36.

Bras, R. L. 1990. *Hydrology*. Reading, MA: Addison-Wesley.

Cunderlik, J. M., S. W. Fleming, R. W. Jenkinson, M. Thiemann, N. Kouwen, and M. Quick. 2013. "Integrating Logistical and Technical Criteria into a Multiteam, Competitive Watershed Model Ranking Procedure." *ASCE Journal of Hydrologic Engineering* (18): 641–54.

Foley, J. A., R. DeFries, G. P. Asner, C. Barford, G. Bonan, S. R. Carpenter, F. S. Chapin, M. T. Coe, G. C. Daily, H. K. Gibbs, J. H. Helkowski, T. Holloway, E. A. Howard, C. J. Kucharik, C. Monfreda, J. A. Patz, I. C. Prentice, N. Ramankutty, and P. K. Snyder. 2005. "Global Consequences of Land Use." *Science* (309): 570–74.

Fleming, S. W., D. R. Bourdin, D. Campbell, R. B. Stull, and T. Gardner. 2015. "Development and Operational Testing of a Super-Ensemble Artificial Intelligence Flood-Forecast Model for a Pacific Northwest River." *Journal of the American Water Resources Association* (51): 502–12.

Freeze, R. A., and R. L. Harlan. 1969. "Blueprint for a Physically-Based, Digitally-Simulated Hydrologic Response Model." *Journal of Hydrology* (9): 237–58.

Levy, S. 2011. "The A.I. Revolution." *Wired*, January, 88–89.

Perkins, T. R., T. C. Pagano, and D. C. Garen. 2009. "Innovative Operational Seasonal Water Supply Forecasting Technologies." *Journal of Soil and Water Conservation* (64) 15–17.

Singh, V. P., and D. A. Woolhiser. 2002. "Mathematical Modeling of Watershed Hydrology." *ASCE Journal of Hydrologic Engineering* (7): 270–92.

Von Neumann, J. 1995. "Method in the Physical Sciences." In *The Neumann Compendium, World Series in 20th Century Mathematics, Vol. 1*, edited by F. Bródy and T. Vámos, 628. Singapore: World Scientific Publishing.

#### CHAPTER 10. EPILOGUE

Böhmelt, T., T. Bernauer, H. Buhaug, N. P. Gleditsch, T. Tribaldos, and G. Wischnath. 2014. "Demand, Supply, and Restraint: Determinants of Domestic Water Conflict and Cooperation." *Global Environmental Change* (29): 337–48.

Crutzen, P. J. 2002. "Geology of Mankind." *Nature* (415): 23.

Fleming, S. W. In press. "Demand Modulation of Water Scarcity Sensitivities to Secular Climatic Variation: Theoretical Insights from a Computational Maquette." *Hydrological Sciences Journal.*

Gleick, P. H. 2000. "Water and Wars." *Eos, Transaction of the American Geophysical Union* (81): 532–33.

Maupin, M. A., J. F. Kenny, S. S. Hutson, J. K. Lovelace, N. L. Barber, and K. S. Linsey. 2014. "Estimated Use of Water in the United States in 2010: USGS Circular 1405." Reston, VA: US Geological Survey.

Murtaugh, P. A., and M. G. Schlax. 2009. "Reproduction and the Carbon Legacies of Individuals." *Global Environmental Change* (19): 14–20.

Ponting, C. 1993. *A Green History of the World*. New York: Penguin.

Serageldin, I. 2009. "Water Wars? A Talk with Ismail Serageldin." *World Policy Journal* (26): 25–31.

United Nations Development Programme. 2006. *Beyond Scarcity: Power, Poverty and the Global Water Crisis, Human Development Report 2006*. New York: UNDP.

United Nations World Water Assessment Programme. 2015. "The United Nations World Water Assessment Report 2015: Water for a Sustainable World." Paris: UNESCO.

Vörösmarty, C. J., P. Green, J. Salisbury, and R. B. Lammers. 2000. "Global Water Resources: Vulnerability from Climate Change and Population Growth." *Science* (289): 284–88.

Wolf, A. T. 1998. "Conflict and Cooperation along International Waterways." *Water Policy* (1): 251–65.

Wolf, A. T., S. B. Yoffe, and M. Giordano. 2003. "International Waters: Identifying Basins at Risk." *Water Policy* (5): 29–60.

# INDEX